Sentient Entanglements and Ruptures in the Americas

Human-Animal Studies

VOLUME 27

The titles published in this series are listed at *brill.com/has*

Sentient Entanglements and Ruptures in the Americas

Human-Animal Relations in the Amazon, Andes, and Arctic

Edited by

Maggie Bolton and Jan Peter Laurens Loovers

BRILL

LEIDEN | BOSTON

Cover image: Llamas and alpacas in Isluga (Chile) showing the ritual dressing with coloured fleece applied in the *wayñu* marking ceremony. Photo: Penelope Dransart, 1987.

Library of Congress Cataloging-in-Publication Data

Names: Bolton, Maggie, editor. | Laurens Loovers, Jan Peter, editor.
Title: Sentient entanglements and ruptures in the Americas : human-animal relations in the Amazon, Andes, and Arctic / edited by Maggie Bolton and Jan Peter Laurens Loovers.
Description: Boston, Massachusetts : Brill, [2023] | "ISSN 1573-4226" | Includes bibliographical references and index. |
Identifiers: LCCN 2023027347 (print) | LCCN 2023027348 (ebook) | ISBN 9789004679443 (hardback) | ISBN 9789004679450 (ebook)
Subjects: LCSH: Human-animal relationships. | Indigenous peoples–Amazon River Region–Social life and customs. | Indigenous peoples–Andes–Social life and customs. | Indigenous peoples–Arctic regions–Social life and customs.
Classification: LCC QL85 .S464 2023 (print) | LCC QL85 (ebook) | DDC 590.89/970811–dc23/eng/20230713
LC record available at https://lccn.loc.gov/2023027347
LC ebook record available at https://lccn.loc.gov/2023027348

Typeface for the Latin, Greek, and Cyrillic scripts: "Brill". See and download: brill.com/brill-typeface.

ISSN 1573-4226
ISBN 978-90-04-67944-3 (hardback)
ISBN 978-90-04-67945-0 (e-book)

This book is printed on acid-free paper and produced in a sustainable manner.

To our Indigenous Teachers in the Americas

Contents

Acknowledgements

We kindly want to thank the organisers of ASA Decennial Conference 2014 *Anthropology and Enlightenment* in collaboration with the STAR Consortium of the Anthropology Departments of Edinburgh, St Andrews and Aberdeen. We specifically want to thank the participants and discussants in our panel 'Objects, Persons or Property? Revisiting Human-Animal Relations in the Andes, Amazonia and the American Arctic'. This book is a direct result of the inspiring papers and discussions.

We want to acknowledge the staff and postgraduate students in the Department of Anthropology, University of Aberdeen who have heard, read or commented upon early drafts of contributions included in the volume. Special mention is due to Tim Ingold and David Anderson. Their respective ERC grants, held for several years simultaneously at the Department, provided a nurturing environment with many discussions on sentient humans and nonhumans.

We are most grateful for the excellent work and assistance of Brill Publishing at Society & Animal and the Human-Animal Series. We particularly want to acknowledge Kenneth Shapiro, Helena Schoeb. Susan McHugh, Brenda Kaldenbach, Fenja Schulz, and the many others behind the scenes. The anonymous reviewers have been brilliant and each of their reviews has enriched both the respective chapter and the volume overall.

Maggie acknowledges the organisers of the workshop *Humanos y otros animales: Relaciones en Transformación, de la crianza a la predación, en el Sur de Sudamérica,* held in Tilcara, Argentina in September 2018, Lucila Bugallo, Francisco Pazzarelli and Penny Dransart. Three of the contributors to this volume also participated in the Tilcara workshop and gained much from the discussions that took place there.

Peter acknowledges the Arctic Domus team of the ERC Advanced grant (no. 295458) for allowing the space and funding for developing his research with dogs and Gwich'in dog mushers. He further is grateful for Florian Stammler and the Arctic ARK (Academy of Finland – 286074) team for providing the space and intellectual engagement to develop his writing for this book.

All the Indigenous Peoples and the nonhuman animals in the Andes, Arctic and Amazonia who participated in the research that was to lead eventually to the contributions in the volume. To you this volume is dedicated.

A special thank you needs to go out to the human (Steve) and nonhuman animals (Smeagol, Max, Freya and the chickens) of Maggie's household and the humans (Ilia and Charalampos) of Peter's household.

List of Illustrations

Notes on Contributors

David G. Anderson
(UK) is a Canadian anthropologist working with communities across the circumpolar North. He started working in the circumpolar North as a technician at the Ft.McPherson Language Centre in the Gwich'in Settlement Area. His reading and fieldwork for his Masters' and PhD was with Evenki people first in Zabaikal'e and later in Taimyr and the Evenki Autonomous Area. Since that time, he has also worked with communities in Iamal, and in Northern Norway. He has been based at the University of Aberdeen since 2000, where he was appointed Chair in the Anthropology of the North in 2013. He has also been appointed as a UArctic Research Chair. He researches and publishes in a wide number of areas: ethnography, ethnohistory, archaeology, Rangifer genetics, and political ecology. His most recent work (with D. Arzyutov and S. Alymov) has been on the history of Soviet Russian *etnos* theory in *Life Histories of Etnos Theory in Russia and Beyond* (Open Books, 2019) and on a model of Arctic Domestication in a series of articles in anthropological and archaeology journals.

Maggie Bolton
(UK) is a Lecturer in the Department of Anthropology, University of Aberdeen, Scotland. She studied for her doctorate at the University of St. Andrews, and subsequently held a Postdoctoral Fellowship, funded by the ESRC, at the University of Manchester. She has conducted ethnographic research in Sud Lípez province in the Department of Potosí, Bolivia since the mid-1990s. Her interests are human-animal relations, particularly among llama herders, and mining. She edited, jointly with Cathrine Degnen, *Animals and Science: From colonial encounters to the biotech era* (Newcastle, Cambridge Scholars Publishing, 2010), and has authored various journal articles and book chapters.

Penelope Z. Dransart
(UK) was awarded a DPhil in Ethnology at the University of Oxford in 1991. Until 2016, she was Reader in Anthropology at the University of Wales, Lampeter. Currently she is Honorary Reader at the University of Aberdeen. Since 1986, she has conducted research in Isluga, in the highlands of northern Chile, on pastoralism, atmospheric weather events, and ritual practices in which human herders and herd animals participate. Stemming from these themes, she also studies the use of fleece from vicuñas, guanacos, alpacas and llamas in archaeological and ethnographic textiles as well as the cultural contexts in which colours are used in such textiles.

Frédéric Laugrand

(Belgium) is a French-Canadian anthropologist, full Professor at Université catholique de Louvain and Director of the Laboratoire d'études prospectives (LAAP), with research expertise on Indigenous cosmologies, Christianization, and human-animal relationships in the Canadian Arctic and in South Asia. He is co-author (with J. Oosten) of *Inuit Shamanism and Christianity: Transitions and Transformations in the XXth Century* (MQUP, 2010), *Hunters, Predators and Prey. Inuit Perceptions of animals* (Berghahn Books, 2014), *Reverend E.J. Peck and the Inuit, East of Hudson Bay (1876–1919)* (Avataq cultural institute, 2019) and *Inuit, Oblate Missionaries, and Grey Nuns in. the Keewatin, 1865–1965* (MQUP, 2019). With E. Luce, he has published *Pelly Bay 1939–1954, F. Van de Velde photographic codex* (PUL, 2019) and with C. Buijs and K. van Dam, *People, Places, and Practices in the Arctic* (Routledge, 2023). He also edited a series of bilingual books on Indigenous knowledge in Nunavut (see the series *Interviewing Inuit Elders; Inuit Perspectives of the Twentieth Century* and *Memory and History in Nunavut*) and in the Philippines, focusing on the Ibaloy, the Blaan, the Ayta and the Alangan traditions (Presses universitaires de Louvain, 2019: https://pul.uclouvain.be/collection/?collection_id=116).

Jan Peter Laurens Loovers

(UK) is a Dutch anthropologist, Research Fellow at the University of Aberdeen in the *Inuksiutit: Inuit Food Sovereignty in Nunavut* project, working in the Arctic over 15 years with research on various themes including dogs, mining, filmmaking, and Indigenous ways of educating. Most recently, he was the Project Curator for the British Museum's Arctic exhibition and a Research Fellow on a project in Scotland about environmental and social justice in the energy transition at St Andrews University. He is the author of *Reading life with Gwich'in* (Routledge, 2020) and co-edited (with R. Losey and R.P. Wishart) *Dogs in the North* (Routledge, 2018) and with A. Lincoln and J. Cooper the acclaimed book, *Arctic: culture and climate* (Thames & Hudson in collaboration with the British Museum, 2020).

Carlos Emanuel Sautchuk

(Brazil) is Associate Professor at the Department of Anthropology, University of Brasilia, Brazil, where he coordinates the Laboratory of Anthropology of Science and Technique. He is also an associate member of the group Anthropologie de la vie et des représentations du vivant (LAS, Collège de France) and co-director of the Centre for the Anthropology of Technics and Technodiversity (UCL). Since 2004 he has been carrying out ethnographic fieldwork in the Amazon on fishing, nature conservation and fish domestication, focusing primarily on anthropological perspectives on techniques, skills and environment. He has coordinated two research projects on technological transformations and he has edited the book *Técnica e transformação: perspectivas antropológicas* (ABA Publicações, 2017) and published the book *O arpão e o anzol: técnica e pessoa na Amazônia* (Ed. Universidade de Brasília, 2021). He has also produced and written about video and photography in ethnography.

Felipe Vander Velden
(Brazil) is Associate Professor at the Department of Social Sciences, Federal University of São Carlos, Brazil. He holds a PhD in Social Anthropology (State University of Campinas, Brazil, 2010) and a postdoc at the Department of Anthropology (Aarhus University, Denmark). He also is currently Affiliated Researcher of the ERC Research Project BRASILIAE. Indigenous Knowledge in the Making of Science: Historia Naturalis Brasiliae (Leiden University, The Netherlands). He has been developing ethnographic fieldwork among the Karitiana Indigenous People in southwestern Brazilian Amazon since 2003 with focus on human-animal relations, animal husbandry, hunting, and history. He is author of *Inquietas companhias: sobre os animais de criação entre os Karitiana* (São Paulo, Alameda, 2012) and *Joias da floresta: antropologia do tráfico de animais* (São Carlos, Edufscar, 2018) as well as of many articles in anthropological journals.

Robert P. Wishart
(UK) is a Lecturer in the Department of Anthropology at the University of Aberdeen, UK. His recent research includes ethnographic, ethnohistorical, and history of science research on human-animal relations in the Canadian North; Indigenous-settler political negotiations; impacts of mining, petroleum development, and wildlife management in the Western Canadian sub-Arctic; and food sovereignty and coastal economies in Scotland.

CHAPTER 1

Introduction: Sentient Entanglements and Ruptures in the Amazon, Andes, and Arctic Regions of the Americas

Jan Peter Laurens Loovers and Maggie Bolton

This edited volume has a long history. It started as a panel we convened at the decennial conference of the Association of Social Anthropologists, held in Edinburgh, in 2014. The conference, *Anthropology and Enlightenment*, organized jointly by the anthropology departments of three Scottish universities, took its inspiration from the Scottish Enlightenment and aimed to rekindle the spirit of Enlightenment debate in a manner appropriate to an anthropology of the twenty-first century. The conference organising committee had identified key themes from Enlightenment works that have retained relevance for current anthropological and interdisciplinary thinking. One of these, intriguingly titled *Human/Nature*, was intended to interrogate the current disciplinary emphasis on rejecting any *a priori* division between nature and humanity in favour of focusing on forms of life as emergent within fields of mutually conditioning relations (c.f. Ingold, 2000). This theme resonated with our shared interest in human-animal relations.

The idea for a panel interrogating human-animal relations in three areas of the Americas – the Andes, Amazonia and Arctic – emerged from the synergies and common themes that we found in our respective work on the Andes with llama herders (Bolton) and on the Arctic with dog mushers (Loovers). Like others writing on our ethnographic regions, we also found that we had both looked, for theoretical and conceptual inspiration, to anthropological studies of Amazonia. We thus concluded that a panel bringing together these three regions of the Americas, ostensibly each very different from the others in terms of environment and climate, but with convergences concerning the position of animals in Indigenous lives, might be a fruitful way to further scholarship.

Returning to the theme of humans and nature in the Scottish Enlightenment, Wolloch (2006) describes how Scottish intellectuals of that era engaged with the notion of theriophily, or the 'love of animals.' This philosophical field covered discussions of human nature, the hierarchical positions of humans and animals, and the moral obligations of humans towards beasts. Such contemplations, we argue, have parallels in the questions and ponderings of

 | DOI:10.1163/9789004679450_002

the Indigenous Peoples of the Americas; both are concerned with maintaining proper relations between humans and different kinds of animals. There are, however, important differences, in how these relations are understood and where animals stand in relation to humans, between the Indigenous philosophies we consider and those from the West. The differences become apparent in the chapters we have brought together here, as do the commonalities between the three American regions, to which we alluded above. While themes of personhood and domestication recur across the contributions in diverse ways and degrees, all our contributions address intriguing similarities, what we might term *sentient entanglements and ruptures* between humans and animals.

Before introducing the contributions to this edited volume, we want to locate our work amid trends in the social sciences and philosophy that seek to decentre the human in analyses. We continue with brief discussions of personhood and domestication, two issues to which scholarship on our regions has made significant contributions in anthropology, and which recur persistently in writing on our ethnographic areas. We then demonstrate how these issues, together with our more general theme of entanglements, enter and are developed in the individual essays.

1 Decentring the Human in the Social Sciences

Within the academic social sciences and related disciplines, a recent philosophical shift has seen an analytic decentring of what is human, and a displacement of human beings and their interests alone from being the main objects of social analysis (Carter & Charles, 2011; Wolfe, 2010). While this trend has taken many forms – actor-network theory (Latour, 2005), the animal turn (Ritvo, 2007), multi-species ethnography (Kirksey & Helmreich, 2012), 'new animism' (Harvey, 2013; Matthews, 2016), and the ontological turn (Holbraad & Pedersen, 2017) –, it shares a common rejection of the notion that human beings alone can be volitional agents, and reconsiders the notion of agency as a property of heterogeneous assemblages of humans and nonhumans. Such 'posthumanist' approaches break away from the dichotomies of nature versus culture or nature versus society that have characterized Western thought since at least the seventeenth century. Dismantling these dichotomies has led scholars working in the field of human-animal studies to give serious consideration to animals as sentient beings with volition that co-habit, co-exist, and form companionships, symbioses, or entanglements with a range of other human and nonhuman animals, materials, media and objects. Questions of animal welfare and animal rights, in some ways not unlike the aforementioned theriophilic endeavours of the Scottish Enlightenment, have entered into human-animal studies and multispecies ethnographies (Birke & Hockenhull, 2012; DeMello, 2012; Kirksey & Helmreich, 2012; Kopsina, 2017; Räsänen & Syrjämaa, 2017).[1] However, some manifestations of posthumanism have also received criticism for maintaining and reinforcing inherent colonial attitudes towards animals and marginalized (Indigenous) communities and for relying too heavily on the categories of the (Western) natural sciences (Haraway, 2008; Todd, 2016; Wolfe, 2011, Blaser, 2014).

Nevertheless, many studies have gone to great lengths to address the commonalities and shared lives of humans and animals and to break away from Cartesian dualities. At the core of the ontological turn and new animism, for example, has been a commitment to take other perspectives, notably those of Indigenous Peoples, seriously. Furthermore, in the recent decades there has been an increasing focus on revising theories of domestication to address the entanglement of human and animal lives (Anderson, Loovers, Schroer & Wishart, 2017; Lien, 2015; Lincoln & Loovers 2020; Stepanoff, 2012; Stammler & Takakura, 2010; Swanson, Lien & Ween, 2018).

2 Persons in the Amazon, Andes, and Arctic

Questions of, and challenges to, what constitutes a person have been an issue since Europeans first set foot in the Americas and are present in reports of the earliest colonial encounters between Indigenous Peoples and Europeans. As conquistadors, clerics, imperial officers and settlers came to the American continent and met with Indigenous Peoples, philosophical conundrums concerning the nature of Indigenous Americans and European people (Viveiros de Castro 1998a), animals, animal sentience, and humanity's connection with other beings became apparent. Concerning animals, Indigenous Peoples went to great depths repeatedly to explain that animals were neither subjugated to humans nor considered instinctive automatons inferior to them. On the contrary, Indigenous Peoples spoke about animals as forceful beings with volition that had to be respected to ensure continuity in life.

In the nineteenth century, academics, who obtained information from older travelogues and scientific literature, including accounts written by Europeans

1 See https://www.animalsandsociety.org/human-animal-studies/key-readings-in-human-animal-studies/ for an extensive list of academic literature on human-animal studies. Absent, however, are works by Indigenous scholars (e.g., Deloria, 2001).

who had been in contact with Indigenous peoples across the American continent, began to engage with non-European understandings of animals. Scottish scholars, for example, were prominent in unravelling human and animal boundaries by examining the notion of *totem* (Frazer, 1887, 1890; Lang, 1903, 1905; McLennan, 1869, 1870).[2] Around the same time, in the 1850s and 1870s, American and British anthropologists elaborated the notion of *animism* to further the idea that animals, celestial bodies, stones and other artefacts were considered sentient in Indigenous cosmologies (Schoolcraft, 1851; Tylor, 1920[1871]; see also Harvey, 2013; Sillar, 2009).[3] Other anthropologists turned to stories to illustrate how nonhuman beings, including animals, were interwoven with Indigenous Peoples' lives (Boas, 1888; Speck, 1905, 1915; Levi-Strauss, 1962; Zuidema & Quispe, 1968). Indigenous scholars, too, contributed to these theoretical treatises with references to their own, as well as other, Indigenous communities (Jones, 1861; Tantaquidgeon, 1942). Nevertheless, many of these earlier accounts, purposefully or unintentionally, failed to question the Western dualities of nature and culture, or nature and society.

However, a shift, labelled by some as the *New Animism* (Matthews, 2016) has taken place in recent years and anthropologists have gradually taken Indigenous philosophies (more) seriously (Anderson, 2017; Ingold, 2000, 2013; Kohn, 2013), although, as Todd rightly points out (2016), Indigenous scholars still find themselves omitted in these debates.[4] Flowing out of the discussions of totemism and animism, notions of personhood and perspectival subject positions gained prominence in anthropological thought particularly through ethnographic studies from the Amazon and the circumpolar North (Brightman, 1993; Descola, 2013; Feit, 2004; Hallowell, 1960; Ingold, 2000; Kohn 2013; McClennan, 1975; Viveiros de Castro, 1998a, 1998b; Blaser, 2016).

What we have styled the *New Animism* can be traced to the work of Hallowell with Anishinaabe people in northern North America in the first half of the twentieth century, and in particular his writing on Anishinaabe ontology (Hallowell, 1960). Hallowell took seriously Anishinaabe insistence that both

2 John Ferguson McLennan's *The Worship of Animals and Plants* (1869, 1870) brought the Anishinaabe (Ojibwa) word *ototeman* to the wider public. *Ototeman* implies one's brother-sister kin (Anderson, 2018; Lévi-Strauss, 1962). Sir James George Frazer (1887, 1890) would bring it to the anthropological audience leading to a lively debate between American-based anthropologists (Goldenweiser, 1910; Powell, 1902), French anthropologists (Mauss, 1950) and English anthropologists (Thomas, 1902).

3 Unlike the word *totem*, animism derived from the Latin word *anima* which means life, soul (see also Ingold, 2007).

4 On the relationship between anthropology and Indigenous studies see Hendry and Fritznor (2012).

animals and the land were intentional, and that persons could include both humans and 'other-than-humans'. He understood Anishinaabe social relations to encompass humans, animals, spirits and objects, and saw distinctions between the different entities in terms of a 'graduation of power', rather than Cartesian dualisms. Hallowell's work is significant because it came to influence two important, yet divergent, strands of theory concerning personhood and animals: the perspectivism of Eduardo Viveiros de Castro, emerging from Amazonian ethnography, and the relational approach of Tim Ingold (2000a), which originated with the author's ethnographic work in northern Finland. Both strands, in different ways, have led towards what might be termed an anthropology beyond humanity, originating with scholars dissatisfied with a discipline that stops at relations between humans, and diminishes the relations with other beings.

Viveiros de Castro (1998a, 1998b) noted in Amazonian cosmologies that how one perceives oneself and others depends on one's subject position (perspective) and the body one inhabits. Spirits and animals perceive humans as animals (either as prey or predators, depending on the perceiver) and perceive themselves as humans. His starting point is that animals were not only human in mythological times – an assertion that many Indigenous Peoples in the Americas make – but that they continue to be human in the present. The distinctions between humans and animals thus cannot be reduced to the Western distinction between nature and culture. Humanity becomes the reflexive condition of a subject from its own perspective, while the condition of animality is that of a body seen from an external perspective, from which it is 'other' (Vanzolini & Cerarini, 2014). Ingold's approach is grounded in relations with, and perceptions, of environments (Ingold, 2000a, 2010, 2013). His essay, *A Circumpolar Night's Dream* (Ingold, 2000b), takes Hallowell's work as its starting point. Ingold is interested in finding a way of understanding the 'continuity of relations between human beings and all the other inhabitants of the earth', (Ingold, 2000b. p. 109) that avoids conventional anthropocentrism. This anthropology beyond humanity joins with, and learns from, the humans and animals 'alongside which we carry on our own lives' (Ingold, 2013. p. 21).

The Andes region remained for a long time on the margins of theoretical debates concerning personhood. Although ethnographers have long noted that Indigenous Andeans inhabit an animated landscape (Allen, 1988, Gow and Condori, 1976), where geographical features are powerful sentient beings, ontology did not take centre stage in anthropological writing, and animals tended to be relegated to the realm of the symbolic (Harris, 1994) or to be considered economic resources (Flannery et al., 1989). Nevertheless, a challenge to the Western separation of nature from politics has been posed by the recent entry of nonhuman entities (the earth, mountains and water) into the political arena, on the back of Indigenous social movements. This in turn has led to new directions in anthropological scholarship on the region that have had wider influence. Marisol de la Cadena (2010, 2015) writes that Indigenous Andean *runa* are *not only* human but ontologically are *with* the sentient earth beings that form their environment, together with whom they form their *ayllu*, the Andean collectivity that encompasses place, family and landholding (Salomon 1991). It is a short step to bring animals into this sentient entanglement of humans and features of the landscape. Dransart (2010. p. 91) considers Andean herd animals as 'acting subjects', with inter-agentive relations with the humans who herd them. Llamas and alpacas, like humans, live in kinship groups and are connected to the landscape, from which their ancestors are held to have emerged.

3 Domestication

Domestication may not seem particularly pertinent to studies of the three regions under consideration, when allegedly Amazonians do not have domestic animals aside from dogs – which probably arrived on the American continent from Asia together with humans – and in some places Muscovy ducks (Stahl, 2014). Early accounts spoke either of an Amazonian 'failure to domesticate', almost implying a deficiency on the part of Amazonian peoples, or of a 'lack of raw material or opportunity'. However, more recent writing has highlighted the shared logic that guides appropriate relations between different classes of sentient beings and excludes what Westerners understand as domestication (Stahl, 2014).

The Andes region, by contrast, is worthy of inclusion in any study of human-animal relations in the Americas on the basis that it is the only region in South America where large animals were domesticated before the arrival of Europeans. It is thus unsurprising that domestic camelids (llamas and alpacas) feature strongly in so many ethnographic and archaeological accounts of the region (Flores Ochoa, 1977, 1988, 2000; Flannery, Marcus & Reynolds, 1989; Bolton, 2006a, 2010; Dransart, 2002, 2010; Nielsen, 1997, 1998). Wild and domestic animals are prominent in the daily lives of Andean people (Palacios Ríos, 1988; Flores Ochoa, 1988, 2000) as well as in their mythology (Gow and Condori, 1976; Payne, 2008; Allen, 2011) and form a rich source of metaphor upon which they draw in illustrating or explaining aspects of their own existence (e.g., Harris, 1994).

Dransart (2002) has argued convincingly that domestication did not follow the same course in the Andes as in Old World societies, on which most writing concerning animal domestication has been based. Taking issue with Ingold's influential article, *From Trust to Domination* (Ingold, 1994), that examines human-animal relations in hunter-gatherer and herding societies, Dransart argues that the relations of Andean herders with their camelids has more in common with Ingold's notion of trust, which he associates with hunter-gatherers, than with his characterisation of relations in pastoral societies. Ethnographers working in pastoral societies, even in the Old World, have since come to similar conclusions where they work (Bunn 2010). Florian Stammler and Hiroki Takakura (2010) bring together scholars to address pastoralism in Northern Eurasia and Africa. They adopt the notion of social significance in which domestic animals and humans are in conversation with, and give meaning to, one another. David Anderson and colleagues, based on ethnographic work in the circumpolar North, likewise, have argued for a more nuanced picture of domestication in which elements of trust, co-existence, dominance, and control blend together (Anderson, Loovers, Shroer & Wishart, 2017; Anderson 2014; also, Loovers, Losey, & Wishart, 2018).

In a further point that finds resonance with work on northern societies (e.g., Losey, Wishart, & Loovers, 2018), Dransart (2002) contests the conventional view of archaeologists and anthropologists that domestication was a 'one-off' event. Based on her fieldwork in northern Chile and the testimonies of Isluga herders, she considers 'domestication' to be a product of the daily herding activities that herders undertake for their animals, and which bring each successive generation of animals into a state of domestication (Dransart, 2002, p. 47; also 2010, p. 90). This view converges with examples from reindeer herding in Siberia. Stépanoff (2017, p. 417), for example, comments that, 'Domestication is not a static relationship between species; it is an unpredictable and ongoing transformation of this relationship and of the species themselves', allowing that the process can have both failures and discontinuities. Such convergences suggest a need to open up dialogue between ethnographic regions.

4 The Contributions: Sentient Entanglements and Ruptures

The contributions we have brought together in this volume are spread evenly over the three regions – that is, we have two contributions each from the Arctic, Andes and Amazonia. The first contribution, situated in lowland South America, takes a processual approach to relations between fish and *caboclo* fishermen in the Amazon estuary. Considering fishing techniques of harpooning for *pirarucu* (a large scaly fish who lives in lakes) and hooking for *gurijuba* (a catfish who lives in the sea) respectively, Sautchuk addresses how death (the killing of fish) brings together techniques, materials, fishermen, fish, medium and ethics. Rather than speaking of human and animal or human and nonhuman relations, Sautchuk illustrates how fishing is not a homogeneous activity but is situated in particular life-flows between fish, fishermen, materials, and environment. Whereas the *pirarucu* is perceived as an equal to the fisherman, intentional and possessing foresight, the *gurijuba* is more elusive, and in this case the hooks, rather than the fish, are perceived as animated and intentional. Sautchuk thus opts to look for agency not only beyond the human but also thus beyond the animal. The nuances that Sautchuk highlights in the two ethnographic cases are important to bear in mind; relations are transformed, but not ended, through the act of killing. Death is not an ending but is rather a movement in a wider nexus of relations.

Laugrand's contribution, from the Arctic region, challenges zoological classification and the concept of species – for Inuit, native caribou and introduced reindeer are different types of animal, while for zoologists they are the same. Using oral history, Laugrand explores why Inuit have been resistant to government initiatives to introduce them to reindeer herding. As in Sautchuk's discussion of the death of the *pirarucu*, for Inuit, the caribou's death in the hunt is a particular moment in an ongoing set of relations where proper actions towards the living and dead caribou are essential for the well-being of a hunter and his relatives. Caribou are situated in a nexus that includes the land, weather, materials and other beings that include the Owner or Mother of the caribou and *ijirait*. *Ijirait* are powerful but dangerous human-like spirits, who can shapeshift into caribou and complicate conventional distinctions between human and animal. As Inuit elder Suvaksiuq observes, '*ijirait* are deceased people turn[ed] into caribou'. Caribou, then, constantly move between life and death in a way that introduced reindeer do not.

Our second Amazonian contribution, from Vander Velden, also concerns animal classification, this time among the Karitiana who live in the Brazilian state of Rondônia, close to the Bolivian border. In an analysis that has many parallels with Laugrand's work, Vander Velden draws attention to the multiplicity of beings that the Karitiana identify as *onças* (jaguars). While it may be possible to 'map' some Karitiana *onças* onto zoologically recognized species – jaguars, other members of the Felidae family and domestic dogs – this is not possible for all. Some *onças* combine characteristics of different species, appearing monstrous, while others border on the fantastic. In a further echo of Laugrand's contribution, one of the souls of a dead human transforms into a fearsome jaguar. Vander Velden is concerned to avoid viewing Karitiana 'types' as a cultural variation on a given reality and is instead led to wonder whether any single human culture is capable of making a truly comprehensive classification of the living things of this world. The West, he argues, has selected one approach: that of stabilising categories through the identification of fixed animal species. The Karitiana, by contrast, take a more dynamic approach; they opt for instability, apprehending the world through a focus on actions and potentialities.

In the first of our Andean contributions, Dransart addresses the personhood of animals, or their personification, in Isluga in northern Chile. She starts from the observation that herders in Isluga include their llamas and alpacas (and sheep, if they have any) in ritual practices as co-participants. They are named, along with humans, as 'pilgrims' in the wayñu ceremony, which aims at their revitalisation. The animals subsequently take an active role in making ritual offerings in their daily lives as they conduct 'wandering pilgrimages' through their animate environment and disperse the woollen decorations with which they were dressed in the ritual. This acceptance of animals as both persons and pilgrims, together with their relationship with sentient topographical features of their environment, leads Dransart to review anthropological scholarship on animism. Wary of rigid categorisations of ontologies, she takes up Bird-David's idea of animism as a relational epistemology; humans personify other entities in different ways through diverse socialisations. In Isluga, wayfarers – both humans and herd animals – inhabit a world of personified landforms that subsist in flux and transformation.

Bolton also writes about camelids in Andean society. Her contribution focuses on a disastrous snowfall that killed large numbers of llamas in the southwest of Bolivia in 2002 and focuses on domestication as well as personhood. She takes up Ingold's (2007, 2010) notion of a 'weather world', in which humans and animals are entangled, noting that Andeans speak of some weather events, as they do of landforms, as sentient or animate. Recalling Dransart's (2002, 2010) insight into domestication as an ongoing process, she examines how the snowfall imperilled this relation. Here, Bolton draws inspiration from the spatial approach to domestication set out by Anderson, Loovers, Schroer & Wishart (2017) who write of architectures of domestication; the creation of settings in which humans and animals can cooperate, and which are places desirable to both types of being. The snowfall brings about a transformation in these architectures, whereby convivial places turn hostile, disrupting the field of relations in which humans and camelids coexist.

Finally, returning to the Arctic, Loovers and Wishart consider domestication and turn to working dogs in northern Canada. Taking up the idea of partnership rather than companionship (c.f. Haraway, 2003, 2008), they address how Gwich'in work together with dogs in shared tasks like setting a fish net. Similar to Sautchuk and Laugrand, Loovers and Wishart show that the designation 'dog' encompasses a variety of different and nuanced relations. Rather than speaking about the weather-world, questions of well-being and food are brought to the fore. Here dogs are part of a larger nexus of relations that include fish, crows, the land and other beings.

In conclusion, this edited volume highlights *sentient entanglements* in three regions of the Americas. On putting together the aforementioned panel on human-animal relations across three vastly different regions of the Americas, we did not know what outcome to expect. It is not surprising that our contributions demonstrate commonalities, between the different regions, since all are grounded in, or heavily influenced by, Indigenous peoples' entanglements with animals and other beings.

To tangle derives from the Old Norse *Þongull* (seaweed) or Old Scandinavian *tagilen* (to involve in a demanding situation). While entanglement in human-animal studies has moved away from the understanding of *tagilen*, the seaweed-like entwinement recurs frequently in anthropological literature on human and animal relations. Anderson, in our volume's afterword, picks up this etymological root and delves a little deeper into seaweed through Strathern's idea of partial connections. But, as Anderson points out, sentient engagements with seaweed, as life, involves both struggles and harmony. In other words, sentient entanglements are not always smooth. Collard, building an argument around cougars on Vancouver Island, considers entanglements to be messy and knotted together in ways that are particularly fraught, spatial, invoke materiality, and involve processes of domination and resistance (c.f. Sharp et al., 2000 in Collard, 2012, p. 24) and Nading, writing about mosquitos and diseases in Nicaragua, defines entanglements as 'the unfolding, often incidental attachments and affinities, antagonisms and animosities that bring humans, animals and things into each other's worlds' (2014, p. 29). Like Collard, Nading attributes to entanglements a 'material, spatial, and temporal' component (Nading 2014, p. 29). In *Swamplife*, her ethnography of the Florida Evergaldes, Ogden (2011) investigates how the lives of human hunters and alligators entangle with the landscape and the colonial history of reclamation and Indigenous displacement. She picks up Deleuze and Guattari's notions of refrain in her analysis (Ogden 2011). Refrains are 'the shifting states, the continually becoming and becoming undone' of the aforementioned relationships (Ogden 2011, p. 45; see on refrains also Hastrup, 2018). Her analysis reminds us that refrains, like rhizomes and becomings, or indeed entanglements, are not necessarily rosy and cosy, there can be 'disconcert' (Loovers, Losey, and Wishart, 2018).

Indigenous scholars, albeit not speaking about entanglements per se, have eloquently addressed the connections between humans, animals, spirits, and the land (Borrows, 1997; Cajete, 2000; Chaudhuri & Chaudhuri, 2001; Deloria, 1999, 2001; Kerasote, 2001; Kidwell & Noley, 2001; Kimmener, 2013; Ornelas, 2007; Watts, 2013; Wilcox et al., 2013; Whitt, Roberts, Norman & Grieves, 2002). Ruptures flow out of improper actions towards others, including animals, spirits, the land, the weather. The human-animal relations our contributions describe involve intimate linkage, transformation and things becoming sometimes undone or ruptured. The contributions, in sum, say something more general about relations between humans, animals and the environments in which they live, about different ways of apprehending the world.

References

Allen, C.J. (1988). *The hold life has: Coca and cultural identity in an Andean community.* Smithsonian Institution Press.

Allen C.J. (2011). *Foxboy: Intimacy and aesthetics in Andean stories.* University of Texas Press.

Anderson, D.G. (2017). Humans and animals in northern regions. *Annual Review of Anthropology*, 46, 133–149.

Anderson, D.G., Loovers, J.P.L., Schroer, S.A., & R.P. Wishart. (2017). Architectures of domestication: On emplacing human-animal relations in the north, *Journal of the Royal Anthropological Institute*, 23(2), 398–416.

Birke, L. & J. Hockenhull (eds.). (2012). *Crossing boundaries: Investigating human-animal relationships*, Brill.

Blaser, M. (2014). Ontology and indigeneity: On the political ontology of heterogeneous assemblages. *Cultural Geographies*, 21(1), 49–58.

Boas, F. (1888). *The central Eskimo.* Edited by J.W. Powell. Vol. 6. Government Printing Office/Bureau of Ethnology.

Bolton, M. (2006). Genetic defects or generative prototypes? Competing models for livestock improvement in southern Bolivia. *Journal of the Royal Anthropological Institute*, 12(3), 531–549.

Bolton, M. (2010). Healthy meat and natural animals: Nature, culture and decolonising knowledge in Bolivia. In M. Bolton & C. Degnen (Eds.), *Animals and science: From colonial encounters to the biotech industry* (pp. 169–195). Newcastle, Cambridge Scholars Publications.

Borrows, J. (1997). Living between water and rocks. *The University of Toronto Law Journal*, 47(4), 417–468.

Brightman, R.A. (1993). *Grateful prey: Rock Cree human-animal relationships*, University of California Press.

Bunn, S. (2010). Animal knowledge: Thinking through deer and sheep in Kyrgyzstan, in M. Bolton and C. Degnen (Eds.), *Animals and science: From colonial encounters to the biotech industry*, (pp. 105–125), Cambridge Scholars Publications.

Cajete, G. (2000). *Native science: Natural laws of interdependence*. Clear Light Publishers.

Carter, B. & N. Charles. (2011). *Human and other animals: Critical perspectives*. Palgrave Macmillan.

Chaudhuri, J. & J. Chaudhuri. (2001). The harmony of nature, norms, and production. In *A sacred path: The way of the Muscogee Creeks*, (pp. 95–103, 113–115). UCLA American Indian Studies Centre.

Collard, R.-C. (2012). Cougar-human entanglements and the biopolitical un/making of safe space. *Environment and Planning D: Society and Space*, 30(1): 23–42.

de la Cadena, M. (2010). Indigenous cosmopolitics in the Andes: Conceptual reflections beyond politics. *Cultural Anthropology*, 25(2), 334–370.

de la Cadena, M. (2015). *Earth beings: Ecologies of practice across Andean worlds*, Duke University Press.

Deleuze, G., & F. Guattari. (2003). A thousand plateaus: Capitalism and schizophrenia. Translated by Brian Massumi. Continuum.

Deloria, V. (1999). Sacred places and moral responsibility. In V. Deloria (Ed), *Spirit and reason: The Vine Deloria Jr. Reader*, (pp. 323–338). Fulcrum Publishing.

Deloria, V. (2001). American Indians and the wilderness. In T. Kerasote (Ed.), *Return of the wild: The future of our national lands* (pp. 25–35). Pew Wilderness Centre.

DeMello, M. (2012). *Animals and society: An introduction to human-animal studies*. Columbia University Press.

Descola, P. (2013). *Beyond nature and culture*. Translated by Janet Lloyd. The University of Chicago Press.

Dransart, P. (2002). *Earth, water, fleece and fabric: An ethnography and archaeology of Andean camelid herding*. Routledge.

Dransart, P. (2010). Animals and their possessions: Properties of herd animals in the Andes and Europe. In M. Bolton & C. Degnen (Eds.), *Animals and science: From colonial encounters to the biotech industry* (pp. 84–104). Cambridge Scholars Publications.

Feit, H.A. (2004). James Bay Crees' life projects and politics: Histories of place, animal partners and enduring relationships. In M. Blaser, H. Feit, & G. McRae (Eds.), *The way of development: Indigenous peoples, life projects and globalisation* (pp. 92–110). Zed Books with Canadian International Development Research Centre.

Flannery, K., Marcus, J. & R.G. Reynolds. (1989). *Flocks of the* wamani: *A study of llama herders on the punas of Ayacucho, Peru.* Academic Press.

Flores Ochoa, J.A. (1977). *Pastores de puna: Uywamichiq punarumakuna.* Instituto de Estudios Peruanos.

Flores Ochoa, J.A. (ed.). (1988). *Llamichos y paqocheros.* Editorial Universitaria UNSAAC.

Flores Ochoa, J.A. (ed.). (2000). *Pastoreo altoanino. Realidad, sacralidad y posibilidades.* Editorial Plural/MUSEF.

Frazer, J.G. (1887). *Totemism.* Adam & Charles Black.

Goldenweiser, A. (1910). Totemism: An analytical study. *The Journal of American Folklore,* 23(88), 179–293.

Gow, R. & B. Condori. (1976). *Kay Pacha.* Centro de Estudios Rurales Andinos Bartolomé de Las Casas.

Hallowell, A.I. (1960). Ojibwa ontology, behaviour and world view. In S. Diamond (Ed.), *Culture and history* (pp. 17–49). Columbia University Press.

Haraway, D. (2003). *The companion species manifesto: Dogs, people, and significant otherness.* Prickly Paradigm Press.

Haraway, D. (2008). *When Species Meet.* University of Minnesota Press.

Harris, O. (1994). Condor and Bull: The ambiguities of masculinity in northern Potosí, in P. Gow & P. Harvey (Eds.), *Sex and violence: Issues in representation and experience* (pp. 40–65). Routledge.

Harvey, G. (2013). Introduction. In G. Harvey (Ed.), *The handbook of contemporary animism,* (pp. 1–12). Acumen Publishing Limited.

Hastrup, K. (2018). Dogs among others: Inughuit companions in Northwest Greenland. In R. Losey, R.P. Wishart, & J.P.L. Loovers, *Dogs in the north: Stories of cooperation and co-domestication* (pp. 212–232). Routledge.

Hendry, J. & L. Fitznor (Eds.). (2012). *Anthropologists, Indigenous scholars and the research endeavour: Seeking bridges towards mutual respect.* Routledge.

Holbraad, M. & M. Pedersen. (2017). *The ontological turn: An anthropological exposition,* Cambridge University Press.

Ingold, T. (1994). From trust to domination: An alternative history of human-animal relations. In A. Manning & J. Serpell (Eds.), *Animals and human society: Changing perspectives* (pp. 1–22). Routledge.

Ingold, T. (2000a). *The perception of the environment: Essays on livelihood, dwelling and skill.* Routledge.

Ingold, T. (2000b). A circumpolar night's dream. In *The perception of the environment: Essays on livelihood, dwelling and skill* (pp. 89–110). Routledge.

Ingold, T. (2007). Earth, sky, wind and weather. *Journal of the Royal Anthropological Institute,* 13 (Special Issue), S19–S38.

Ingold, T. (2010). Footprints through the weather-world: Walking, breathing, knowing. *Journal of the Royal Anthropological Institute,* 16 (Special Issue), S121–S139.

Ingold, T. (2013). Anthropology beyond humanity. Edward Westermarck Memorial Lecture, May 2013. *Suomen Antropologi: Journal of the Finnish Anthropological Society*, 38(3), 5–23.

Jones, P. (1861). *History of the Ojebway Indians: With especial reference to their conversion to Christianity*. Bennett.

Kerasote, T. (2001). *Return of the wild: The future of our national lands*. Pew Wilderness Centre.

Kidwell, C.S. & H. Noley. (Eds). (2001). *A Native American theology*. Orbis Books.

Kimmerer, R.W. (2013). *Braiding sweetgrass: Indigenous wisdom, scientific knowledge and the teachings of plants*. Milkweed Editions.

Kirksey, S.E. & S. Helmreich. (2012). The emergence of multispecies ethnography. *Cultural Anthropology*, 25(4), 545–576.

Kopsina, H. (2017). Beyond multispecies ethnography: Engaging with violence and animal rights in anthropology. *Critique of Anthropology*, 37(3), 333–357.

Lang, A. (1903). *Social origins*. Longmans, Green, and Co.

Lang, A. (1905). *The secret of the totem*. Longmans, Green, and Co.

Lévi-Strauss, C. (1962). *Totemism*. Translated by Rodham Needham. Merlin Press.

Latour, B. (2005). *Reassembling the social: An introduction to actor-network theory*. Oxford University Press.

Lien, M.E. (2015). *Becoming salmon: Aquaculture and the domestication of a fish*. University of California Press.

Lincoln, A. & J.P.L. Loovers. (2020). Relations with animals in the circumpolar north. In A. Lincoln, J. Cooper & J.P.L. Loovers (Eds.), *Arctic: Culture and climate* (pp. 44–77). Thames & Hudson in collaboration with the British Museum.

Loovers, J.P.L., Losey, R. & R.P. Wishart. (2018). Conclusion: Dogs in the north. In R. Losey, R.P. Wishart, & J.P.L. Loovers, *Dogs in the north: Stories of cooperation and co-Domestication*, (pp. 278–292). Routledge.

Matthews, M. (2016). *Naamiwan's drum: The story of contested repatriation of Anishinaabe artefacts*. University of Toronto Press.

McClennan, C. (1975). *My old people say: An ethnographic survey of Southern Yukon Territory*. National Museums of Canada.

McLennan, J.F. (1869–1870). The worship of animals and plants. *Fortnightly Review* 4 (1869): 407–27, 562–82 & 7 (1870): 194–206.

Nielsen, A. (1997–8). Tráfico de caravanas en el sur de Bolivia: Observaciones etnográficas e implicancias arqueológicas. *Relaciones de la Sociedad Argentina de Antropología* XXII–XXIII, 139–178.

Ornelas, R. (2007). Understanding sacred lands. *Great Plains Research*, 17, 165–71.

Palacios Ríos, F. (1988). Pastores de llamas y alpacas. In X. Albó (Ed.), *Raíces de América: El Mundo Aymara*. Alianza Editorial – UNESCO.

Payne, J. (2000). *She-calf and other Quechua folk tales*. University of New Mexico Press.

Powell, J.W. (1902). An American view on totemism, *Man*, 2: 101–106.

Räsänen, T. & T. Syrjämaa. (2017). *Shared lives of humans and animals: Animal agency in the global north.* Routledge.

Ritvo, H. (2007). On the animal turn, *Daedalus* 136(4), 118–122.

Salomon, F. (1991). Introductory essay: The Huarochiri Manuscript. In F. Salomon and G. Urioste (Eds.), *The Huarochiri Manuscript: A testament of ancient and colonial Andean religion*, (pp. 1–42). University of Texas Press.

Schoolcraft, H.R. (1851). *The American Indians: Their history, condition, and prospect.* George H. Derby and Co.

Sillar, B. (2009). The social agency of things: Animism and materiality in the Andes. *Cambridge Archaeological Journal* 19(3), 369–79.

Speck, F.G. (1909). *Ethnology of the Yuchi Indians.* The University Museum, University of Pennsylvania.

Stahl, P.W. (2014). Perspectival ontology and animal non-domestication in the Amazon Basin. In S. Rostain (Ed.), *Antes de Orellana. Actas del 3er encuentro internacional de arqueología Amazónica* (pp. 221–231), Institut Français des Études Andines.

Stammler, F. & H. Takakura. (2010). *Good to eat, good to live with: Nomads and animals in northern Eurasia and Africa.* Tohoku University (Northeast Asian Study Series 11).

Stépanoff, C. (2012). Human-animal 'joint commitment' in a reindeer herding system. *Hau: Journal of Ethnography,* 2, 287–312.

Stépanoff, C. (2017). Comment on Anderson et al., 'Architectures of domestication', *Journal of the Royal Anthropological Institute,* 23(2), 417–418.

Swanson, H.A., Lien, M.E., & G.B. Ween. (2018). *Domestication gone wild: politics and practices of multispecies relations.* Duke University Press.

Tantaguidgeon, G. (1942). *A study of Delaware Indian medicine practice and folk belief.* Pennsylvania Historical Commission.

Thomas, N.W. (1902). An American view of totemism. *Man,* 2, 115–118.

Todd, Z. (2016). An Indigenous feminist's take on the ontological turn: 'Ontology' is just another word for colonialism. *Journal of Historical Sociology,* 29(1), 4–22.

Tylor, E.B. (1920 [1871]). *Primitive culture: Researches into the development of mythology, philosophy, religion, language, art, and custom.* John Murray.

Vanzolini, M. & P. Cesarini. (2014). *Perspectivism.* Oxford Bibliographies. DOI: 10.1093/obo/9780199766567-0083.

Viveiros de Castro, E. (1998a). Les pronoms cosmologiques et le perspectivisme amérindien. In E. Allier (Ed.), *Gille Deleuze: Une vie philosophique,* (pp. 429–62). Les Empêcheurs de Penser en Rond.

Viveiros de Castro, E. (1998b). Cosmological deixis and Amerindian perspectivism. *Journal of the Royal Anthropological Institute,* 4(3), 469–488.

Watts, V. (2013). Indigenous place-thought and agency amongst humans and non-humans (First Woman and Sky Woman go on a European world tour!). *Decolonisation: Indigeneity, Education & Society,* 2(1), 20–34.

Whitt, L.A., Roberts, M., Norman, W. & V Grieves. (2002). Indigenous perspectives. In D. Jamieson (Ed), *A Companion to Environmental Philosophy* (pp. 3–20). Blackwell Publishing.

Wilcox, A.C., Harper, S.L., Edge, V.L., Landman, K., Houle, K., Ford, J.D., & the Rigolet Community Government. (2013). The land enriches the soul: On climatic and environmental change, affect, and emotional health and well-being in Rogolet, Nunatsiavut, Canada. *Emotion, Space and Society*, 6, 14–24.

Wolfe, C. (2010). Moving forward, kicking back: The animal turn. *Postmedieval: A Journal of Medieval Cultural Studies*, 2(1), 1–12.

Wolloch, N. (2006). The status of animals in Scottish enlightenment philosophy. *Journal of Scottish Philosophy*, 4(1), 63–82.

Zuidema, R.T. & U. Quispe. (1968). A visit to God: The account and interpretation of a religious experience in the Peruvian community of Choque-Huarcaya. *Anthropologica*, Vol. x: 22–39.

CHAPTER 2

Moral Gestures: Forms of Life and Forms of Death in Amazonian Waters

Carlos Emanuel Sautchuk

Abstract

This chapter examines the gestures of killing involved in two main modalities of artisanal fishing in the Amazon River estuary – using harpoons on the region's lakes and longlines on the coast. It seeks to make sense of the inevitable centrality of the death of fish without reducing this fact to ethnocentric dilemmas of utilitarianism or animal rights, for example. Also avoiding recourse to broader, generic categories such as (multi)species or humans and nonhuman animals, the different meanings and moralities of killing are considered within the technicities of these two fishing methods. In focusing its approach on the shared semiotic materialities of these two activities, the chapter also explores the meanings of the death of fishermen in both contexts. It thus ethnographically demonstrates that their forms of life correspond to particular forms of death, including the gestures of killing and their dilemmas.

Keywords

Amazon – fish – technique – death – nonhuman animals

1 Introduction

Addressing debates about relations between human and nonhuman animals from the perspective of encounters between fish and fishermen necessarily involves the question of death. It can be assumed generally, even from a non-utilitarian point of view, that fishing is about capturing fish and changing them into either food or a commodity. Therefore, relations with the living animal and the way its flesh is distributed and consumed should not be downplayed. Yet, the act of killing remains unavoidable – for the fish, for the fisherman and thus for the anthropologist – since it constitutes the teleology and condition for these relations. This leads us to ask how we are to make sense of the

 | DOI:10.1163/9789004679450_003

centrality to fishing of the death of fish without reducing fishermen-fish relations to a utilitarian perspective.

This question, which is at the core of my analysis in this article, has broader implications. Death further nuances anthropological efforts at approaching so-called multi-species or human/nonhuman animal relations. However, it is not only anthropologists who are affected by this encounter with death. On two occasions, when I presented my research on fishermen living in the Amazon Estuary to different audiences (Sautchuk, 2020), I screened a short video (Sautchuk, 2007) by two groups of fishermen from Vila Sucuriju, a village located in the Brazilian Amazon. The video shows the capture of two different fish: the *pirarucu* (the arapaima fish, *Arapaima gigas*) harpooned in a lake area, and the *gurijuba* (a yellow catfish, *Arius parkeri*), on board an artisanal motorboat along the Amazon sea coast. Scenes of the capture and death of fish caused diverse reactions in the audiences, with opinions changing in different directions.

In a meeting with state officials and advisors in Brasilia concerning the demands of the fishermen's association for the recognition of *pirarucu* harpooning as Brazilian cultural heritage, those scenes caused a setback (although this was later reversed). Some of those in the meeting who had hitherto been sympathetic to the fishermen's claims, presented in terms of the valorisation of Amazonian traditional knowledge, reacted negatively to the scenes of the death of fish. This was especially accentuated in the case of the *pirarucu*, one of the world's largest scaled fish. On another occasion, in a meeting to deliberate whether to maintain fishermen's access to the lake area where they used to work, located within a protected area, the same video produced a positive effect. Impressed with the ingenuity and skill required to capture such a large and quick-witted fish, most of those present began to value the fishermen and their sophisticated knowledge and skill. A generally sympathetic sentiment concerning their access to the area thus emerged from an originally unfavourable context.

Notwithstanding the socio-political complexities and particularities of each situation, it can be said that these reactions to the death scenes point in opposite but correlated directions for considering animals in urban modernity. In fact, the admiration and even romanticization of the fishermen's skill and courage coexist with repulsion over the violence of the death of the fish, even among people who enjoy eating them.[1] As historian Keith Thomas (1984)

1 The tension between repulsion and attraction in response to the fish's death are widely explored in the visual arts and literature aimed at the urban and literate public. Recently, the art exhibition and the film *The Fish* caused unease by showing Brazilian fishermen

showed, modern sensibilities regarding the preservation and exploitation of plants and animals have emerged in tandem with the idea of nature. Even if we acknowledge the complexity of views on animals in Western philosophy (e.g., Agamben, 2004; Simondon, 2004), the weight of the nature–culture binary is still felt in the meanings attributed to animals. Yet, approaching this relation from the two modernist poles – in Thomas's terms, of conquest and preservation – may prevent us from making sense of other peoples' relations with animals.

Descola (1998) reminds us that Western perceptions have been characterized by a notion of nature as a dimension external to the human. In the case of animal death, this has clear consequences:

> Where nature does not exist as an autonomous sphere, relations with animals cannot but be different than ours, and the question of killing an animal can only be put in terms very different than our own. This is what a detour through the Amazon could help establish. (Descola, 1998, p. 25, translated by the author)

Descola, referring here to Indigenous sensibilities in the Amazon, draws a contrast with Western animal protection movements such as deep ecology. He suggests that the presence of death makes it even harder to understand other ways of relating with animals. This difficulty is precisely what makes situations of death good to think with, at least from the standpoint of anthropological concerns over the meanings of animals. Similar claims about the meaning of death have been made for hunters from multiple sites, prompting interesting debates that challenge modern binaries such as subject–object or nature–culture, and which speak directly or indirectly to some recent key debates in contemporary anthropology (for instance, Ingold, 2000, 2013; Willerslev, 2004; Viveiros de Castro, 1998; Descola, 2013). Here, I look at Amazonian fishermen from the perspective of current anthropological debates concerning relations between humans and nonhuman animals. Firstly, these are people (the so-called *caboclos* or mestizos; see Harris, 1998 and Wagley, 1953) manifestly influenced by Amerindian (and Afro-Brazilian) cosmologies and practices.

affectionately caressing fish as they died asphyxiated in their arms (Farago, 2007). This ambivalence – admiration/repulsion, pride/guilt – is present in diverse forms in the fictional literature on fishing, including classics like Moby Dick and The Old Man and the Sea. Indeed, contemporary recreational fishing attempts to resolve this dilemma by valorising the experience of capture and the duel, but without the death of the fish, in a transformation similar to sportification (Franklin, 1996).

However, additionally, and in broader terms, those who relate directly to animals and to their death do not put this question in the same vexed and passionate terms as we do in Western urban contexts – since, in general, we do not directly kill animals ourselves, but only eat some and coexist with others (Descola, 1998).

On this last point, Marvin (2006) has shown how killing, and particularly hunting animals in contemporary Europe, can show an understanding of death that is different from the modern way of thinking about it (see also Hell, 1996). Praet (2014) offers an interesting take on the contrast between animism, where life is more restricted, and nature conservation, where it is considered as an extensive property or phenomenon. Conservationism, informed by a western way of thinking, considers diversity in life, but not of death (Praet, 2014, p. 165).

The current general response to the act of killing animals reveals a dose of collective contradiction, undoubtedly caused by the genuine unease over the mass industrialisation of animal production and the corresponding alienation between urban consumers and farms. It is common for people to like eating meat but not wish to participate in, observe, or even know about, the death of the being on their plate, as I heard explicitly in the case of fishing. Donna Haraway has perceptively shown the contradictions between this posture and the problem of not treating death as a phenomenon of life. For her it is crucial to care about the death of animals in the contemporary world, but, 'how to take it seriously is far from obvious' (Haraway 2008, p. 335). She adds that it is a misstep to oppose the two alternatives – animal rights and human profit. Haraway argues that a morality of killing relations can hardly be encountered in a humanist epistemology, and neither does the egalitarian premise of equivalence help us. Avoiding both the idealist and symmetricist trap, we need to consider death within the flows of emotions, techniques, and practices wherein a 'nonanthropomorphic sensibility that is accountable to irreducible differences' (Haraway 2008, p. 90) can flourish.

I would like to take a step in this direction by comparing the meanings of death in the cases of two fish targeted by lake and coastal Amazonian fishing – the *pirarucu* and the *gurijuba*, respectively. Based on these differences, I suggest that fish and fisherman – their lives and deaths – should be understood less as a multi-species interaction or by drawing on such charged categories as 'humans' and 'nonhuman animals'. To explore more closely the ethnographic meaning of the gestures of killing, it is better to frame the question in terms of the joint emergence of these 'animate beings', based on the imperative of going 'beyond humanity' (Ingold, 2013), which also means going beyond the notion of animal.

In dealing with the death of animals, we need to keep in mind Haraway's methodological warning not to demonize instrumentality or its use. This means considering beings and their acts within a shared semiotic materiality, including the moralities internal to 'unequal and ontologically multiple instrumental relationships' (Haraway, 2008, p. 73). Indeed, we can also observe that among authors who have argued in favour of an ethnographic approach to the relation between life forms and forms of life (Helmreich, 2011), Pitrou (2015) advocates a pragmatic approach, pointing precisely to the fact that by examining techniques we can perceive multiple coordinating actions. In this sense, by affirming the centrality of life, the ethnography is unavoidably faced with the meaning of death. In other words, it becomes inevitable to articulate life forms and forms of life with the meanings and moralities present in different forms of death.

2 Fish

The situations presented here refer to Vila Sucuriju, a riverside community to the north of the Amazon estuary in the state of Amapá, Brazil. The approximately five hundred inhabitants live from fishing and are divided into two groups; coastal fishermen dedicated to line fishing along the coast in motorized boats, and lake fishermen (or *laguistas*) who use harpoons to catch fish in the lake region. The caught fish circulate in networks of local exchange (gift), but are sold via the *aviamento* (debt bondage) system, generating flour and other items. Agriculture and animal breeding are lesser activities due to the regular tidal flooding. Certain forms of subsistence hunting are also practiced, such as hunting for capybaras, while fishing takes place in sustainable manners, including through agreements reached with the Brazilian government (Sautchuk, 2017, 2019). The relationship with certain animals like piranhas is ambiguous, permeated by predation and a kind of commensality (Sautchuk, 2016). In the community, exotic animals, like chickens and pigs, are bred, which become food on special occasions. Amongst the companion animals kept are cats and dogs, but also native fauna such as parrots and monkeys. These are not eaten since they become members of the family, protected from attacks by jaguars and other predators. For example, the death of a giant otter, the companion animal of one of the fishermen who was harpooned by accident while submerged, was much grieved and lamented. Among this wide variety of forms of life and forms of death present in relations with animals, I concentrate here on those existing in lake and coastal fishing.

It should be noted from the outset that a general aversion towards animal deaths is not prevalent among the lake and coastal fishermen of Vila Sucuriju. Children are initiated early on (Sautchuk, 2005, 2015) into the processes whereby fish (as well as other animals) are transformed into *embiara* – the name given to the flesh that results from the capture, treatment, and distribution of animals, which are not done without due caution and attention to the consequences. However, killing and causing suffering in vain are not tolerated, and capturing skills are developed at the same time as the apprenticeship of moral dispositions associated with the control and appropriate use of violence, not just against animals (Sautchuk, 2015).

On the lakes, as on the coast, the act of killing, although routine, is a highly relevant fishing task, and a prestigious skill among fishermen. In both places, lethal gestures share remarkable similarities, since both involve *cacetar*, that is, to club the fish's head after it has been captured. However, a closer look indicates that not just the meaning of the act but its dynamic is particular to each situation: on the lakes, it is carried out with circumspection and deference; on the coast, it is a joyful and euphoric moment for the boat crew.

The two kinds of lethal gestures are closely associated with attributes particular to the two kinds of fish. Noteworthy for its ingenuity and intentionality, the *pirarucu* is described in terms of caution, foresight, calculation, and intelligence. The lake fisherman can only capture it when the fish *agrees* to be captured – even if capture is considered as a kind of duel-like game, it is only successful when the animal wilfully gives itself up. Like hunting by Indigenous Peoples across the Circumpolar North more generally (Ingold, 2000; Willerslev, 2004; Nadasdy, 2007), here the fish concurs with its own death. The fatal gesture is therefore the final stroke in a movement to which the fish itself, as subject of its own death, converges (see Sautchuk, 2019).

The *gurijuba*, in its turn, is not personified. It is treated as a treasure – it is also called 'golden fish' – and coastal fishing is often compared to gold mining. As locals say, the fisherman *comes across the fish* as he prospects the sea, in what is considered a largely unpredictable and fortuitous event, whose causes, normally stemming from the local popular Catholicism, transcend human-fish interaction. There is also an analogy with agriculture, in the use of terms such as 'seeding' and 'harvesting' to designate fishing tasks.

This comparative effort can be taken further to unveil a series of differences. These are found, for instance, in situations where the fish pose a threat to the fishermen. The *gurijuba* presents a risk after it has bitten the hook; but any sting is attributed to carelessness or a contingency inherent in the activity, rather than to the fish's intention to harm. The *pirarucu*, in its turn, can present an added risk to the fisherman's capacity to fish, as it may engender *panema* (Wagley, 1953; DaMatta, 1973), a state of inability to capture, caused by the inappropriate distribution of meat, or careless treatment of the animal and its leftovers – two actions that give rise to deep concern.

During consumption, the *gurijuba's* eggs, and fat from the head, are sources of strength for the fishermen, as they increase the virile capacities associated with fishing and sex. The most important aspect of *pirarucu* consumption, in its turn, is less the transmission of properties coextensive to fish and fisherman through food than the circulation of meat as a form of exchange. Accordingly, if coastal fishing's cuisine values non-commercial parts of the *gurijuba* (eggs, head, fin), on the lakes, the *pirarucu*'s noblest parts are consumed locally regardless of their commercial value, thus shoring up a direct relationship between the harpooner and *his* fish (or, more precisely, his *embiara*).

But it is in the modes of capture that the most remarkable particularities lie. To catch the *pirarucu*, the fish's master (or *dono*, that is, the spirit that controls and protects the fish) must release the *pirarucu* for the fisherman. As for the *gurijuba*, a good capture is a consequence of divine intervention – when God *looks the fisherman in the face*, He grants him a felicitous expedition.

It must be noted that, since these two specialized groups of fishermen live in the same village, they share a similar, as it were, 'socioeconomic and cultural background' – historical and economic trajectory, memory, and local identity. But how can we move forward in our analysis without attributing the different meanings of the two fish to an abstract construction, to qualities intrinsic to the animals or to adaptations to the environments (lake versus sea) where they are predated? Multiple ethnographic counterexamples (such as objects or animals found in both situations) would easily invalidate any attempt at finding a symbolic, biological, or environmental causality behind these differences.

3 Beyond Nonhuman Animals, towards Techniques

It seems fundamental, first of all, to exclude any perspective on humans and animals that would regard them as being fundamentally different (Ingold, 2000). Such an approach would lead to views on their relations based on analogies or causality. In the first case, animals and humans would appear as parameters for each other. In this sense, both anthropomorphic perspectives, where animals are viewed according to human realities, and behaviourist and sociobiological approaches, which think of humans from the standpoint of a background of animality, unfold from the same cleavage. According to Ingold (1988, 2013), the first strategy may be criticized for its anthropocentrism, since it fixates human singularity as the comparative criterion for conceptions of animals found in different cultures. The second is, in its turn, ethnocentric, since the idea of biological and behavioural continuity between humans and animals stems from Western notions of evolution and human nature.

Up to a point, it can be said that even in anthropology the question of human–animal relations has been framed through such binaries. In his review of the topic, for instance, Shanklin (1985, p. 379) found that few authors before him had tried to avoid the opposition between functional and symbolic perspectives on animals. In fact, it seems that, for a while, the main question was how to describe relationship patterns between two more or less fixed poles – environment-nature-animality and society-culture-humanity – from which the various approaches got their legitimacy precisely by rivalling one another (Guille-Escuret, 1989). This may be an effect, within anthropology, of the more general way whereby modern thought has conceived of oppositions between humans and animals (cf. Ingold, 1988, p. 95; Descola & Pálsson, 1996).

In this sense, any ethnographic approach to human-animal relations that does not avoid the dualistic assumption (be it in terms of analogy or causality) will distort them, or at least impoverish them in certain senses. Among the alternatives available, which include multispecies ethnographies and, obviously, the work of Tim Ingold, I would like to draw firstly on the French ethnology of technique and two of its chief claims. André-Georges Haudricourt called into question the human-animal dichotomy by advocating a homology of relations (rather than of beings) between humans (or social) and vegetables and animals (or technique). Commenting on Haudricourt's (1968) seminal article, *'Domestication of animals, cultivation of plants and human relations'*, Descola (2013, p. 106) praised his defence of the homology between the principles underlying human–nonhuman relations, concluding that there is neither simply a projection of human relations onto animals nor an extension, to humans, of relations between animals. We should think instead in terms of common principles orienting behaviour towards various beings and about taking actions and gestures as a way of making sense of relations between humans and animals (cf. Rival, 2012; Ferret, 2012).

Another student of Mauss dedicated to technique, André Leroi-Gourhan (1993) also criticized Cartesian binaries and rejected the de-materialising stance of the human sciences, which led to a 'denial of any connection between the human and the rest of the living world' (1993, p. 400). He claimed that the intelligence–instinct opposition does not overlap with its human–animal counterpart, and proposed instead that what matters when characterising beings are the modalities of grouping, that is, the various technical behaviours that, through their rhythm, make up collectivities. This inscribes in the same analytical plane phenomena that would have been otherwise distinguished as social, zoological, or technical (cf. Stiegler, 1998; Ingold, 1999). Leroi-Gourhan and Haudricourt were early critics of views of animals and humans as belonging to separate domains. Instead, they underscored a common ground where the permeability of qualities and characteristics is predicated on the kinds of relations entertained with each other. Based on these assumptions, they put forth a series of tenets for an empirical approach that would substantiate a processual (or operative) perspective, which avoids thinking in terms either of species or of human–nonhuman animal relations.

Among the contemporary authors who have advanced similar claims, sometimes influenced by these early approaches on techniques, are Phillipe Descola and Tim Ingold. Though presenting major divergences, both put forth perspectives in contemporary anthropology that underscore relations with the environment. Descola's (2013) structural phenomenology proposes that some relations between humans, animals and artifacts are inter-subjective and emerge according to ontological principles that organize interactions between them. Tim Ingold, in his turn, agrees that animal and human persons constitute their identities and particularities reciprocally, by means of their interactions. He does not, however, grant analytical primacy to any structural organisational dimension, but looks instead at practical interactions. Referring to hunter-gatherers, Ingold claimed that:

> (…) the constitutive quality of their world is not intersubjectivity but *interagentivity*. To speak of the forest as a parent is not, then, to model object relations in terms of primary intersubjectivity, but to recognize that at root, the constitutive quality of intimate relations with nonhuman and human components of the environment is one and the same. (Ingold, 2000, p. 47)

It must be underlined that Ingold's proposal is centred not on energetic flows, but on the phenomenological and semiotic foundations of practical interactions (Reed, 1988, p. 113; see also Bateson, 2000, p. 459). The notion of affordance (Gibson 1979) is key here: it indicates that the environment is predicated on the possibilities for interaction presented to a given animal. Somewhat similar to the notion of the *Umwelt* (Uexküll, 1982), which conceives of animal behaviour from the standpoint of its 'own world', the notion of affordance suggests that the physical environment has objective, albeit particular, meanings, since these are based on a given animal's possibilities for perception and action. This is also true for perception between animated beings: in other words, the meaning of an animal is predicated on the interactive possibilities that another animal establishes with it, in a dynamic and situated process (cf. Reed, 1988; Sebeok, 1988).

This perspective thus rejects the notion that all meaning is an effect of projection, anthropomorphism or symbolic elaborations, since interactions involve characteristics that are proper to other beings and objects (Gibson, 1979). Another approach that reinforces this point comes from semiotics, in authors such as Sebeok, for whom the meaning of animals is neither inherent to them (as in a substantialist ethology), nor arbitrarily bestowed upon them by humans, but comes instead from their reciprocal interactions: 'To appreciate what counts as an animal for man and in what ways, finally requires a concentrated semiotic enquiry (...)'. In order to do so, he goes on to argue, '... it seems promising to consider the many and varied circumstances under which man may encounter animals.' (Sebeok, 1988, p. 68). Examples include relations of predation (in both directions), partnership, parasitism, entertainment, and sports.

Tim Ingold has long been at the forefront of efforts in anthropology to revise perceptions of nonhuman animals based on these concerns. Most recently, he has sought to displace two tenets of Western thought – the unity of the individual and the society-nature divide – by proposing that persons, including animals, 'are not beings that move, they *are* their movements' (Ingold, 2011, p. 168). The animal therefore can only be understood in terms of a field of relations which is itself part of a movement (Ingold, 2000, p. 95–8).

In what follows, I return to the case of Amazonian fishing in the light of these debates. Here, I draw ethnographically not on such large categories as human and nonhuman animals, but on the movements of fishermen and fish, and on shared relations between them. In other words, I focus on the technical processes of capture and killing *per se*, as they have meaning (Pfaffenberger, 2001). After all, discourses and behaviours are grounded in the practices of lake and sea fishing. Moreover, it becomes necessary to find parameters for establishing a contrast between the two kinds of fish, according to the relative (that is, non-essentialist) statutes that they embody in each environment. In other words, it is not about comparing the fish as such, but in terms of the role each plays in flows of techniques on the lake and in the sea. It is germane to evoke here Rival's (1996) proposition that hunting with different weapons evokes different meanings of an animal and establishes different kinds of relation with it.

To move forward, it is thus useful to understand the gestures of killing and their morality within the technicities prevalent in lake and coastal fishing. I thus present fishing techniques in these two environments in the Amazon estuary as relations that ontogenetically give meaning to fish, fishermen, and other beings. This focus on technique follows two fundamental orientations. Firstly, it encompasses multiple relations, behaviours, objects, and environments, which involve interactions that cannot be reduced to instrumental ends. Here, technique is synonymous with relation, and concerns after all, not outcomes, needs or utility, but action itself (Mauss, 2009; Leroi-Gourhan, 1993). Secondly, the centrality of gestures and knowing-doing evokes Ingold's (2000) perspective, where skill does not come from the active subject (or organism-person), and neither is it previous or external to the activity, but emerges out of practical interactions (Ingold 2011).

4 *Laguista* and *Pirarucu*

Pirarucu fishing on the lakes is carried out with harpoons, on two-person paddle canoes. This kind of fishing foregrounds the harpooner's encompassing role, since it involves a face-to-face encounter with the fish, based on confrontation, cunning, and seduction. Like other lake animals (caimans, anacondas, piranhas, monkeys), the *pirarucu* is endowed with a particular and unique personality, characterized by active interaction with men. Its capacity to flee, deceive and even beat the *laguista* (or, conversely, the fact that it may surrender itself on certain occasions) correlates with the fisherman's reciprocal engagement in this game of interpretation and skill, which he only joins as a subject bearing his technical apparatus: harpoon, canoe, and pilot.

In this modality of fishing, everything happens in the path between the lake-bottom (*fundo*) – that is, the entire liquid medium, not just its deepest layer – and the *buiado* – the aerial medium adjacent to the water. For this reason, the *flor d'água* – the boundary between the aquatic and aerial media – is a reciprocal communicative interface, a plane of display and dissimulation of actions and intentions. It is there that the *laguistas* avoid to inscribe their presence and orient themselves according to the set of visual signs produced by the fish, which require sophisticated decoding capacity. As much as the harpooner pays attention (*repare*), he only *sees* the *pirarucu* when the fish presents itself to him – the sign he identifies is always a fish's signal (*aviso do peixe*). Once the *laguista* locates the *pirarucu*, the hunt is on. The goal is to find the best position from which to throw the harpoon without being noticed by the fish. If, however, the latter notices the fisherman's presence, all procedures of stealth are suddenly suspended. The situation then changes: the fish flees, deploying multiple stratagems, and pursuit requires ingenuity. In any case, the opportunity (*enseio*) – that is, the proper moment to harpoon – is made possible by the fish itself. It is only when this happens that the *harpooner* must fling the harpoon, and then pull the fish in by the rope attached to the weapon that penetrates its flesh. Hence the value *laguistas* attribute to the key aptitude to *não se afobar*, that is, not to anticipate the fish's signal – for this would amount to cutting off the dialogue with the animal, to compel him.

As if the *pirarucu*'s cunning (*fineza*) and intentionality were not enough, harpooning also involves further nuance, since the fish occupies centre stage in the key events that lead to its own capture. The flight of the weapon may last as long as two seconds, a time interval during which the fish also moves. Harpooning is, in fact, a convergence between the harpoon and the fish at a future point in time. Its success is therefore also predicated on the fish's behaviour. Moreover, as the harpoon meets the *pirarucu*, it only penetrates the fish if the latter gets a fright (*se espante*), thus bending its body and opening up its scales. If this happens too early, the fish flees, and if it happens too late, the animal remains stiff, and the harpoon is not able to pierce the scales.

Therefore, even if the harpooner's ability to handle the harpoon is the condition for his existence as such, this ability alone does not lead to capture. Dexterous harpooning is understood as a given, universal disposition among harpooners – it is therefore not with skill alone that capture lies, but in the relation between this skill and the *pirarucu*. In other words, the fish anticipates the skill involved in harpooning, and its intention is highly relevant for the *laguista*'s personhood.

Moreover, the will of the *pirarucu* (and of its master) is decisive not only for capture. *Pirarucu* fishing is intimately related to the fisherman's personhood when it comes to distributing the flesh – a process that must follow a series of restrictions and prescriptions closely associated with kin networks. Several precautions must be taken to avoid certain dangers that may harm a fisherman's productivity by undermining the *pirarucu's* trust in him.

Engaged in a relation that resembles both duelling and seduction, the *pirarucu* is considered a person with communicatory and intellectual abilities comparable with those of *laguistas*, in due proportion with their bodily particularities. The transformation of humans into fish is in fact a common subject of conversation in the lake area, and is conceptualized as a sort of change in outfit which has, however, serious consequences.[2] It must be kept in mind, nonetheless, that the harpooner also thinks of himself in terms of this particular idiom of predation and intersubjectivity; thus, it is by following the particular interactive parameters of this modality of predation that both – *laguista* and *pirarucu* – become persons. And it is in these dramatic terms that fish death is understood on the lakes.

2 Interpersonal relations between humans and animals in the context of hunting have been a common topic, and this issue has been further elaborated through notions of perspectivism (Viveiros de Castro, 1998), animism (Descola, 2013), and interagentivity (Ingold, 2000).

5 Coastal Fisherman and *Gurijuba*

It is hard to think of anything less believable than a coastal fisherman becoming a *gurijuba*. By the sea, relations between fisherman and fish are very different than on the lakes. Coastal fishing involves eight- to twelve-meter motorboats carrying three or more crew members. Its main piece of equipment is the *espinhel* – a longline made up of hundreds of hooks, fixed close to the sea bottom by anchors and floats – and in the rainy season it also uses a large gill net positioned in the tidal water. Either way, the entity that organizes work on the sea is the boat (*barco*), a term encompassing the vessel itself, the engine, fishing gear, and the boat crew. Its identity is compounded by the boat name, its particular characteristics (such as speed, and how it handles the sea waves), its stories, the quality of its equipment and its crew. It is the boat that dries up, turns around, floats (*buia*), touches the bottom – in sum, that brings the fish (*traz*). It is individualized as a 'reliable boat', or a boat that 'sways too often', or one that is 'tough', and so forth.

The hook, which is personified in multiple ways, is an indispensable partner in coastal fishing, since it is the entity that accomplishes capture on the sea bottom. However, it has the nasty habit of unexpectedly betraying the fisherman by hooking him. This ambivalence stems from the fact that this technical object's animated character is predicated on its traffic between the bottom (where it is subjected to water flows) and the dynamics of onboard tasks - between the water current and the fisherman's hands. The gill net is based on the same principle of relationship with the flow of the tide. In this case, the boat is moored next to the net and the work is organized by tide times. As is the case with the hook, *the net fishes*.

This dynamic is encompassed by a broader canvass of qualities and interactions among other elements involved in this modality of fishing. Faced with a constantly changing hydric environment, the boat's components function in close connection with each other, including by means of ropes. As a result, coastal fishermen value corporeal qualities associated with physical interactions between bodies, following general parameters of size, strength, and resistance. Unlike in lake fishing, the objects and beings (fishermen and fish included) are qualified according to their capacity for kinetic interaction rather than perception and communication. Each one of them occupies a position in a peculiar kind of chain, ordered according to principles of resistance and movement.

In this context, the *gurijuba* is regarded as a kind of recondite wealth along the lines of gold that is panned, where chance plays an important role. However, in coastal fishing, although the resource being sought is, as gold, submerged in the texture of an unexplored space, the fish are not static. To put it differently, their position is solidary to space – this is however a space-time, a space-flow, which alters the characteristics of places and the distance between them according to the intensive tide cycles in the Amazon estuary. In contrast with gold panning, which is done by revolving the terrain, here it is the boat that is fixed by its equipment (the longline or the gill net) and must resist the hydric temporality of currents. The boat therefore lets this fluid space-time pass through it, bringing with it the desired resources, or not, as the case may be – the catfish moves on the sea bottom, led by the tidal currents.

In this case, the unpredictability of capture has a different character than on the lakes, since it is associated less with the intentionality of fish than with tidal movements. This is significant because, besides being a condition for communication (and trade), the water flow is a pillar of local cosmogony and a principle guiding time. The entities responsible for ordering this universe – God and the Virgin Mary – act precisely in the imponderability of bodily interactions. While they may save a fisherman from a fishhook's *pulo* (leap) or save the boat from a tidal wave (*pororoca*), they may (or not) lead the *gurijuba* to the longline or the net placed on the sea bottom.

Although the *gurijuba* may take on characteristics similar to those of the fishermen, these relate less to their direct perception and communication than to their capacity to (re)act physically to other beings or objects present in these technical systems. This quality is generally, but differentially, distributed, and prominence is granted to the boat (the encompassing entity), the engine and the fishhook, in addition to the tides. The *gurijuba* moves along with the currents and is caught by the hook (or the net) even when it does not take the bait. But this does not diminish the fish – in fact, it shows its value after it has been hooked, by tugging the line and therefore altering the rhythm of onboard tasks. In sum, fish and fishermen are located in distinct planes of a common system of action that encompasses and fixates the parameters for its own existence. Just like the fisherman, the *gurijuba's* meaning relates to the place that it occupies in the technicity of coastal fishing.

6 Shared Semiotic Materialities of Killing … and Dying

The question of death requires a processual approach: it is not just an end, but a transformation, whereby something is closed (and not just the animal's life) while something else is open (and not just the creation of foodstuff or commodity). Even if the animal's death is indeed central, the fishermen on the lakes and on the sea do not look at it from a utilitarian or animal-centred perspective (be the animal conceived as organism or as person). The gesture of clubbing is just one amid a cycle of other actions, an act that puts an end to one thing (or to some properties of a life form) and initiates another. In their relations with the fishermen, the meanings of fish involve a perspective on death that cannot be reduced to one conception of what is life and what is an animal. The attachments that end (or begin) with these lethal gestures belong to distinct orders, even if they are reciprocally implicated.

Pirarucu and harpooner constitute each other as autonomous and equivalent beings, engaged as they are in reciprocal processes of perception and action. To predate the *pirarucu* involves engaging, entering the perceptual world, the *umwelt* of this 'other', and the reverence that presides over the lethal gesture reflects not only a consideration for the surrender of an equal rival. It is also a sublime and solemn moment for the *laguista* since it demarcates the onset of concerns over the catch (common in Amazonian hunting at large). As in hunting societies, this moment marks the absorption of bodily alterity, whereby the *laguista* comes to perceive himself as a distinctive being in a context where the qualities and characteristics of personhood are distributed across humans and animals (cf. Descola, 2013, p. 397). The sequence of blows inflicted on the fish is ritualized, involving increased rhythm and decreased intensity: it begins with firm, fatal blows, and ends with soft, almost affective pats. This gesture, which seems almost musical, produces a very distinctive sound on the bony skull of the *pirarucu* that propagates across the lakes the harpooner's pride in his capture, as well as his self-assigned responsibility and care for that death. There is no true harpooner – a paragon of masculinity – prior to or without the continuity of this ontogenetic percussion.

On the coast, the exultation that accompanies the deadly act reflects the good fortune of seeing the *golden fish* – and not another, valueless, animal – emerge from the bottom, captured by the hooks. Like almost all other gestures made on board, clubbing the fish combines agility and the strength required for dealing with other bodies amidst the sea ripples. Done with precision, it gives way to the actions that follow of collecting the line and avoiding accidents involving the hooks or fish stings. In this case, the skill involves rapidly arresting the fish's movements without jeopardising its most valued product, the swim bladder. It is by terminating its movements that the fisherman incorporates the fish into the boat as a product; this gesture is yet more evidence of his integration with onboard activities. The death of a fish thus condenses the synchronisation between the boat's two components: schedule (tide) and tasks (onboard activities). To kill a *gurijuba* is a fundamental act for sustaining the onboard dynamic, or the articulation of bodies encompassing the fishermen. This is however separate from capture, which is attributed to the hooks

FIGURE 2.1 *Pirarucu* and harpooner
PHOTO: C. SAUTCHUK

as well as to divine designs and is directly implicated in stalling the movements associated with the longline and the tides.

But if the technicities of these two forms of fishing involve fishermen and fish in a shared semiotic materiality, it is necessary to consider the deaths of both. In her reflection on the ethical dimension of animal killing, Haraway underlines the value in thinking about the different forms of death, as well

FIGURE 2.2 Fisherman and *gurijuba*
PHOTO: C. SAUTCHUK

as their reversibility: 'I do not think we can nurture living until we get better at facing killing. But also get better at dying instead of killing' (Haraway, 2008, p. 81). It is not a question here, of course, of presuming a symmetry or equivalence between deaths, in part because while the death of the fish is the condition and the objective of these two activities, the death of the fisherman is a possibility and an accident. Even so, if we wish to go beyond an ethnocentric critique, we need to avoid units of analysis based on the individual or the species and ask ourselves too about the form in which the fishermen die. After all, the technical relations establish the morality of the engagement between the lives and deaths of fish and fishermen – and pervade the very emergence of these as actors (Latour, 2002). Put otherwise, if the life forms involved in these techniques imply different forms of life (Pitrou, 2015), they also present singular forms of death, both for fish and for fishermen.

The occurrence (or the possibility) of the death of a fisherman, even if rare, is very significant. I cite two cases of death during fishing that manifest different aspects of the kinds of relations that prevail on the lakes and on the coast, especially when it comes to the meaning of *fundo* (bottom). Both events involve apparent death by drowning: a coastal fisherman was taken by the tide, and a senior *laguista* died 'under a couple of inches of water'. Both men died after penetrating the hydric domain, but the situations were very different. The first was checking on the net while walking on the beach, when the tide rose violently and dragged him into the sea before the boat could be floated to try to save him. The other met a weird death in a shallow area of a lake, which was attributed to the evil deed (*malineza*) of some master – a consequence of the deceased's inappropriate acts towards the animals.

In both cases, the explanations for the death of fishermen bear connections with those for the death of fish. On the coast, the water dynamics typify the landscape by exercising their power over objects and fishermen. The fatality occurred because the measure and intensity of relations between the task and schedule were not appropriately predicted or managed, not just by the deceased fisherman, but by the boat itself, as it is the chief subject of this relationship. Since the forces in play are much bigger than those of the objects, fisherman or fish, any unfavourable connection can kill or cause harm to them. The lakes, on the other hand, are peopled by intentionalities and traversed by interactions with other beings that are not always visible. Death in this case was understood as a consequence of the late fisherman's inappropriate actions, which provokes the reaction of other beings.

In sum, we could state that the *pirarucu* stands for the *laguista* and harpoon as the *gurijuba* stands for the fisherman and hook. The meanings of fish are therefore predicated on their relations with humans and objects. By this I do not mean that a fisherman's death has the same value or importance as the death of a fish, but that both share a semiotic materiality present in a technique. In this sense, I suggest that the relationship between fisherman and fish may be interpreted based on the flow of relations whence both emerge, thus avoiding a substantialist or pronominal perspective. As advocated by Ingold, we could think in terms of verbs: 'Humans, baboons and reindeer do not *exist*, but humaning, babooning and reindeering *occur*—they are ways of carrying on' (Ingold, 2013, p. 21).

In its radicality, death makes evident how the existences of fishermen and fish must be viewed from the perspective of the actions that institute them as such, including those properties and qualities that give meaning to their end. This is why death is a different phenomenon on the lakes and on the coast, since *pirarucu*-harpooner and *gurijuba*-fisherman becomings are different, albeit related, grounded as they are on perceptual-communicational or kinetic capacities, respectively. We could say that fatality and the designs of God and the Saints are the main explanations for the death of fishermen and *gurijuba* on the coast – it always happens as it should be, even if it is tragic. On the lakes, everything is based on intentionality, and we never know for sure what and who is at stake. Everything could have been different if the intentions had been otherwise. Simplifying, we could say that fisherman and *gurijuba* do not die as a consequence of someone's action, but as a result of designs and non-personnel forces whose logics are unfathomable. *Pirarucu* and harpooner usually die as a result of the relationship between different and uncertain intentionalities.

In this sense, utilitarian or symbolic perspectives are not enough for making sense of death during fishing. In each modality of fishing, the transformations engendered by the gesture of 'killing fish' (*matar peixe*) transcend the animal and the human, and only make sense in terms of the 'instrumental intra-actions' (Haraway, 2008, p. 71). By approaching the idea of a shared semiotic materiality through an ethnographic focus on technique, the objective is precisely to emphasize that this sharing should not be understood primarily as an interaction but as an ontogenesis, not as an outcome but as a condition. A technique, in the relational sense adopted here, implies the emergence or transformation of the actors involved in it, thus creating its own moralities (Latour, 2002).

These gestures of killing are, therefore, best understood from a perspective that reaches beyond a symmetric and narrow (and generic) focus on (multi) species or human and nonhuman animal relations, to encompass fish, fishermen, objects, and the environment. It seems more appropriate to ask whether we can consider ontogenesis *qua* technical relations, whereby fishing and the beings, things and processes implicated in it are simultaneously engendered. In activities, such as fishing in particular, we are led to affirm that focusing on forms of death – including killing, dying and its moralities – is central to understanding forms of life.

References

Animal Studies Group (Ed.). (2006). *Killing animals.* University of Illinois Press.

Agamben, G. (2004). *The open: Man and animal.* Stanford University Press.

Coupaye, L. (2016). Yams as vernacular methodology? Approaching vital processes through technical processes. *Les actes de colloques du musée du quai Branly Jacques Chirac.* Retrieved September 21 2019, from: http://journals.openedition.org/actesbranly/673.

DaMatta, R. (1973). Panema: uma tentativa de análise estrutural. In *Ensaios de antropologia estrutural* (pp. 63–92). Vozes.

Descola, P. (1998). Estrutura ou sentimento: a relação com o animal na Amazônia. *Mana, 4*(1), 23–45.

Descola, P. (2013). *Beyond nature and culture.* University of Chicago Press.

Descola, P. & G. Pálsson (Eds). (1996). *Nature and society: Anthropological perspectives.* Routledge.

Farago, J. (2017, January 26). When predator and prey are one: A Brazilian artist's fish tale. *The New York Times.*

Franklin, A. (1996). On fox-hunting and angling: Norbert Elias and the «Sportisation» Process. *Journal of Historical Sociology, 9*(4), 432–456.

Ferret, C. (2014). Towards an anthropology of action: From pastoral techniques to modes of action. *Journal of Material Culture, 19*(3), 279–302.

Gibson, J.J. (1979). *Thẹ ecological approach to visual perception.* Houghton Mifflin.

Guille-Escuret, G. (1989). *Les sociétés et leurs natures.* Armand Colin.

Haraway, D. J. (2008). *When species meet.* University of Minnesota Press.

Harris, M. (1998). 'What it means to be Caboclo': Some critical notes on the construction of Amazonian Caboclo society as an anthropological object. *Critique of Anthropology, 18*(1), 83–95.

Haudricourt, A.G. (1969). Domestication of animals, cultivation of plants and human relations. *Social Science Information, 8,* 163–172.

Hell, B. (1996). Enraged hunters: The domain of the wild in northwestern Europe. In P. Descola & G. Pálsson (Eds.), *Nature and society: Anthropological perspectives* (pp. 205–218). Routledge.

Helmreich, S. (2011). What was life? Answers from three limit biologies. *Critical Inquiry, 37*(4), 671–696.

Ingold, T. (1988). Introduction. In T. Ingold (Ed.), *What is an animal?* (pp. 1–16). Unwin Hyman.

Ingold, T. (1999). 'Tools for the hand, language for the face': An appreciation of Leroi-Gourhan's gesture and speech. *Studies in the History and Philosophy of Biological and Biomedical Science, 30,* 411–453.

Ingold, T. (2000). *The perception of the environment: Essays on livelihood, dwelling and skill.* Routledge.

Ingold, T. (2011). *Being alive: Essays on movement, knowledge and description.* Routledge.

Ingold, T. (2013). Anthropology beyond humanity. *Journal of the Finnish Anthropological Society, 38*(3), 5–23.

Latour, B. (2002). Morality and technology. *Theory, Culture & Society, 19*(5–6), 247–260.

Leroi-Gourhan, A. (1993 [1964]) *Gesture and speech.* MIT Press.

Marvin, G. (2010). Challenging animals: Project and process in hunting. In *Nature and culture.* Earthscan Publications.

Mauss, M. (2009). *Techniques, technology and civilisation.* Durkheim Press, Berghahn Books.

Nadasdy, P. (2007). The gift in the animal: The ontology of hunting and human–animal sociality. *American Ethnologist, 34*(1), 25–43.

Pitrou, P. (2015) Life as a process of making in the Mixe Highlands (Oaxaca, Mexico): Towards a 'general pragmatics' of life. *Journal of the Royal Anthropological Institute,* 21(1), 86–105.

Praet, I. (2014). Animal conceptions in animism and conservation. In G. Marvin and S. McHugh (Eds.), *Handbook of human-animal studies* (pp. 154–167). Routledge.

Pfaffenberger, B. (2001). Symbols do not create meanings - activities do: Or, why symbolic anthropology needs the anthropology of technology. In B. Schiffer (Ed.), *Anthropological perspectives on technology* (pp. 77–86). University of New Mexico Press.

Reed, E. (1988). The affordances of the animate environment: Social science from the ecological point of view. In T. Ingold (Ed.), *What is an animal?* Unwin Hyman.

Rival, L. (1996). Blowpipes and spears: The social significance of Huaorani technological choices. In P. Descola & G. Pálsson (Eds.). *Nature and society: Anthropological perspectives* (pp. 155–174). Routledge.

Rival, L. (2012). Comment on John Knight 'The anonymity of the hunt'. *Current Anthropology,* 53(3), 38–9.

Sautchuk, C.E. (2005). Laguistas et pescadores: l'apprentissage de la pêche dans une région côtière de l'Amazonie (Vila Sucuriju, Brésil). *Techniques et culture,* 45, 61–186.

Sautchuk, C.E. (2007). *The harpoon and the hook.* Documentary video, 6 min. (available at http://www.youtube.com/watch?v=_mGqXjz4Sqo. (accessed 4/20/2019).

Sautchuk, C.E. (2012). Cine-weapon: the poiesis of filming and fishing. *Vibrant,* 9(2), 406–430.

Sautchuk, C.E. (2015). Aprendizagem como gênese: Prática, skill e individuação. *Horizontes Antropológicos,* 21(44), 109–39.

Sautchuk, C.E. (2016). Eating (with) piranhas: untamed approaches to domestication. *Vibrant: Virtual Brazilian Anthropology,* 13(2), 38–57.

Sautchuk, C.E. (2017). Matar e manter: Conservação ambiental como transformação técnica. In C. E. Sautchuk (Ed.), *Técnica e transformação: Perspectivas antropológicas* (pp. 183–212). ABA Publicações.

Sautchuk, C.E. (2019). The pirarucu net: Artefact, animism and the technical object. *Journal of Material Culture*, 24(2), 176–193.

Sautchuk, C.E. (2020). *O arpão e o anzol: técnica e pessoa no estuário do Amazonas*. Editora Universidade de Brasília.

Sebeok, T. (1988). 'Animal' in biological and semiotic perspective. In T. Ingold (Ed.), *What is an animal?* Unwin Hyman.

Shanklin, E. (1985). Sustenance and symbol: Anthropological studies of domesticated animals. *Annual Review of Anthropology*, 14(1), 375–403.

Simondon, G. (2004). *Deux leçons sur l'animal et l'homme*. Ellipses.

Stiegler, B. (1998). *Techniques and time, 1: The fault of Epimetheus*. Stanford University Press.

Thomas, K. (1984). *Man and the natural world: Changing attitudes in England, 1500–1800*. Penguin Books.

Uexküll, J. von. (1982). The theory of meaning. *Semiotica*, 42(1): 25–82.

Viveiros de Castro, E. (1998). Cosmological deixis and Amerindian perspectivism. *Journal of the Royal Anthropological Institute*, 4(3), 469–488.

Wagley, C. (1953). *Amazon town: A study of man in the tropics*. Macmillan.

Willerslev, R. (2004). Not animal, not not-animal: Hunting, imitation and empathetic knowledge among the Siberian Yukaghirs. *Journal of the Royal Anthropological* Institute (N.S.), 10, 629–652.

CHAPTER 3

'We Want to Kill Caribou, Not to Live with Them': Inuit Cosmology and Resistance to Herding

Frédéric Laugrand

Abstract

This paper explores the cosmological foundation of Inuit rejection of herding and the primal relationships of migratory caribou and local spirits called *ijirait*. The study is drawing on classical ethnographical materials and oral sources collected in the Canadian eastern Arctic. Information was recorded from Inuit Elders between 1997 and 2011. The question of herding animals, specifically caribou and reindeer, emerged in the early 1920s when Saami herders were brought to Baffin Island. For various reasons, this adventure turned into a major failure. A closer look at Inuit perception of caribou illuminates the ontological foundations of Inuit's historical and continued resistance to and resentment of herding. In the Canadian Arctic, now Nunavut, Inuit remain attached to hunting the caribou, which they refer to as 'the lice of the earth'. They connect the caribou to a master spirit (*inua*) or God, and often associate caribou with the deceased and mountain spirits (*ijirait*). With this cosmology, and given that hunting and herding involve divergent ideas about animals' agency and human–animal relationships, transforming Inuit hunters of caribou into herders seems a very difficult task. For Inuit, caribou cannot be mastered or domesticated, and they cannot be 'products'. They are 'subjects' to be eaten and respected. Their paths were object of many rules in order to let the caribou go their proper way. Disrespect would trigger starvation. Hunting caribou was – and to a large extent still is – considered the only way to preserve them.

Keywords

Inuit – caribou – *ijirait* – hunting – herding – shamanism – cosmology – master spirit – animism

 | DOI:10.1163/9789004679450_004

1 Introduction

In the 1920s, Kaj Birket-Smith (1929) stated that 'the Sub-arctic Indians, the Eskimos, the Siberian tribes, and the Lapps, present certain fundamental points of resemblance in culture' (p. 244), and all these groups share a common resource, the caribou or the reindeer. Since then, the idea that the Polar peoples or the Indigenous Peoples of the Circumpolar North all live on this animal has taken root, as has the idea that caribou and reindeer are, in fact, a single animal (*Rangifer tarandus*). Thus, reindeer and caribou were (and still are) considered to belong to the same single species (e.g., Ingold, 1980, p. 17, quoting Banfield, 1961, p. 103). Inuit, however, provide a contrasting perspective. Not only do they consider caribou and reindeer to be very different animals, but Inuit were, and remain, rather reluctant to become reindeer herders, with some exceptions. This reluctance was evident, for example, during the Amadjuak experiment of the early 1920s, when the Canadian government and the Hudson's Bay Company brought Saami families and a herd of reindeer from Norway to Baffin Island to train Inuit hunters as reindeer herders. The experiment was a failure (see Laugrand & Oosten, 2015). Much earlier in Alaska and later in the Mackenzie Delta, similar initiatives were hardly more fruitful, even if the contexts were quite different (see Plattet & Lincoln, 2014; Conaty & Lloyd, 2004). Inuit remained very marginal in these activities and left the work to the Saami and the Chukchi.[1]

This paper explores the cosmological foundation of Inuit rejection of herding and the primal relationships of migratory caribou and local spirits called *ijirait*; these relationships have sociological implications insofar as they can help explain Inuit aversion to adopting reindeer herding. I focus on how Inuit relate to the caribou and why they consider this animal so different from the reindeer.[2] I will argue that although caribou are a good prey, they are always potentially ambiguous. Caribou are possibly owned by a master spirit or by God and this has important consequences. They have to be hunted to avoid depletion, and are a major source of food. Second, they are also strongly connected to the land and closely associated with the *ijirait*, the mountain spirits, as well as the deceased and ancestors. For all these reasons, they are thought

1 In Siberia, some groups, such as the Chukchi, hold reindeer herding and hunting at the same time, which shows that herding and hunting can coexist (see Descola, 2005, p. 499–515; Stépanoff et al., 2017). Among the Chukchi, reindeer domestication involves a whole way of life and a particular connection to the domestic fire (Vaté, 2007, p. 228).

2 Given space limitations, I will not explore the reindeer-herding complex here. I refer the reader to the huge body of literature on this topic: see Anderson (2000), Vitebsky (2005), etc.

to be different from the reindeer; they cannot be mastered or domesticated, and they cannot be 'products'. They are 'subjects' to be eaten. The sources used for this paper encompass classical ethnographical documents and information recorded from Inuit Elders between 1997 and 2011. As such, the paper is not aimed to contribute to 'new animism' that has recently developed in the context of our contemporary ecological crisis (see Harvey, 2017, 2019); rather, it deals with the category of animism as redefined by Philippe Descola (2005) as a mode of identification and as cosmology.

2 Following the 'the Lice of the Earth' and the Owner of the Caribou

For Inuit, caribou hunting has always been, and remains, essential for survival. Inuit greatly appreciate the meat, skins, and bones of the caribou (for more details, see Tuck, 1954; Laugrand & Oosten, 2014). Most Inuit hunt caribou, *tuktu*, in the spring and autumn, and some groups in the Kivalliq, such as the Ahiarmiut, depended almost completely on it for survival. Simon Tookoome (1999) from Qamanittuaq described the processes and rationale of caribou hunts:

> The caribou used to gather in a very large herd to migrate ... We would walk among them and pick out the fat ones. We would kill them carefully without frightening the herd. If we did not hunt caribou, they would multiply and there would become too many of them. Then they would suffer and starve until the herds were the right size again.
>
> Inuit would never set up camp in the path of the caribou. We wanted them to go their proper way. If someone put up a tent near a path, the caribou would change their route.
>
> When Inuit kill an animal in the traditional way, they give it a drink of water or, if there is no water, they melt snow with their mouths. The water is very helpful. Water eases the animal's suffering even after it is dead.
>
> We also removed the sinews before we cracked the leg bones. This was also to ease the dead animal's pain. The spirit of the animal was grateful. We respected the animals and the animals allowed us to catch them. After we killed a caribou, the tradition was to point the skin towards your home or tent. It sent more caribou that way. (p. 28–29)

Inuit see themselves as followers of the caribou. They follow the rules they have learned from their Elders and do their best to avoid animal suffering. They show their respect to the game and to the meat of the animals, which usually cannot be sold but only shared. The practice of orienting the skin towards the hunter's home is an interesting procedure. Today, with permanent houses, we can speculate that when Inuit place a caribou skull on their house or shack, it works like a sign. By this act they may still intend to indicate to the caribou the path of the good hunter's house (see Figures 3.1 and 3.2). Especially with the caribou, their path and tracks – their *tugluniit* according to Eva Murjungniq from Arviat – were very much respected by hunters (Oosten & Laugrand 2010, p. 67).

Inuit hunters were also supposed to listen carefully caribou demands which were expressed vocally. Eva Murjungniq recalled that their snorting sound is called *kalaqtuq* but that it cannot be considered as a song. Yet, her husband Job Murjungniq from Arviat suggested that caribou might express themselves to the hunter, 'There was a person who was going after a bull, and that bull said, 'Don't spear me right now. Spear me in a while, because your wife will not be able to have caribou meat that has fat in it when summer comes.' I have heard

FIGURE 3.1 Indicating the hunter's house to the caribou. Photographs taken in Arviat, Nunavut, Canada
PHOTO: F. LAUGRAND, 2011

FIGURE 3.2 Indicating the hunter's house to the caribou. Photographs taken in Arviat, Nunavut, Canada
PHOTO: F. LAUGRAND, 2011

about a bull saying that.' (Oosten & Laugrand, 2010, p. 69). Eva Murjungniq stated that a lack of respect of animals would lead to starvation,

> We all understand that we have to respect caribou. We all know that. I heard that any animal, whether it is a caribou or not, if you treat it badly, could get back at you. ... If we are not going to eat the animal, or if you make fun of it, then you are going to run out of food and go hungry. There will be no animals, nothing to kill. Even here in the community there are no caribou sometimes because we are not treating them with respect. That's what it means. We try not to make fun of any living thing or treat them badly, but still we are doing this without really being aware that we are doing that. Caribou are disappearing. Sometimes we have no meat to eat, no food to eat. When we respect animals we have plenty of food. The animals are to be respected. (Oosten & Laugrand, 2010, p. 67)

In Inuit cosmology, the creation of the caribou is unclear, and there are several different origin stories. The first one says that the caribou came out of the earth. Franz Boas (1907, p. 536–37) relates that most people lived on small game, but one man would bring in a caribou every day. His fellow Inuit discovered that he would lift up a large flat stone and the caribou would jump out from underneath. One day, the stone was turned and many caribou kept jumping out – that is how they scattered all over the earth. The caribou coming from the earth evokes the idea of a multitude. This idea is preserved in the term for caribou in the shamanic language – *kumarjuait* – which means 'big lice'. The term indicates a shift in scale as it denotes the caribou as lice crawling the earth.

The connection of caribou with the earth has always been and remains strong. Knud Rasmussen (1929, p. 202–203) recounted that *silaat*, giant animals associated with rain and mist, that turn into caribou, are called the children of the earth. Thus, *silaat* are connected to *nuna*, the earth, but also to Sila (the universe).

The second origin tradition connects the caribou and the walrus. It says that these two animals were created as a pair by the mother of the caribou, who is sometimes equated with the owner of the sea mammals. Rasmussen (1929, p. 67) noted that, among the Iglulik, some people believe the sea woman also owns the caribou, whereas others think the caribou have their own owner. Rasmussen (1929) provided an account of the origin of the caribou as related by Orulo, the wife of the Aava, a shaman. The 'Mother of the Caribou', 'the one with whom the caribou are', is described as an old woman who went up inland to make the caribou: 'She made their skins from her own breeches, and changed their original tusks into antlers, kicking them on the forehead which made the hollow one can still see in the caribou. And she went to live among them' (Rasmussen, 1929, p. 67–68).

Boas (1888/1964) also recounted that in Cumberland Sound, among the Akudnirmiut, the walrus and the caribou were thought to be created by a woman from her breeches and boots. He observed, 'It is very probable that this woman was Sedna, as the Eskimo affirm that the observances referring to walrus and deer are commanded by Sedna and as the first tradition accounts for her dislike of the deer' (Boas, 1888/1964, p. 588). In this variation, the caribou were made out of the boots and the walrus out of the breeches of a woman. This story exists in many variants, but, importantly, it always establishes a connection between the walrus and the caribou. This connection may be evoked by the fact that walrus, like caribou, are gregarious and can often be found in multitudes basking in the sun.

Captain Lyon (1824, p. 361) distinguished between the sea woman and Pukimna, the owner of the caribou. Boas (1888/1964, p. 588) could not find any trace of this tradition, but Captain Low (1906) stated, 'There is a goddess of the land-animals called Pukimna, who appears to be closely identified with Nuliayok [or Nuliajuk, the goddess of sea mammals], and may be the same personage under a different name' (p. 169). She may also be related to the Pinga, a spirit located in the sky by the Padlirmiut in the Kivalliq region (see Laugrand & Oosten, 2008, p. 25–32). Rasmussen (1929, p. 113) also provides the name Pakitsumanga for the mother of the caribou, and explains that she can provide *angakkuit*, shamans, with *qaumaniq*, shamanic vision. But Rasmussen (1929, p. 70) also stated that most Iglulingmiut considered the mother of the caribou and the mother of sea game to be the same being.

Reverend E.J. Peck, who helped Boas translate some of his ethnographical text recorded in shamanic language, was told that the caribou were owned by 'Tikkertserktok, the goddess of the land (*noonaut innunga*)' (Laugrand, Oosten & Trudel, 2006, p. 369). Peck assumed that she was even more powerful than Sedna. Thus, the mother of the caribou was a powerful being. She could be distinguished from the sea woman but also identified with her. She created the caribou and was their owner. She required gifts, soles and lines, at the start of the caribou hunting season; otherwise, she would make people sick (Laugrand, Oosten & Trudel., 2006, p. 369).

The Iglulingmiut also had to make a gift when they started caribou hunting. Rasmussen (1929) observed, 'People hunting caribou in kayaks on a river or lake must, while hunting, lay out a piece of sealskin under a stone as a sacrifice to Tugtut Igfianut, the Mother of the Caribou' (p. 195). The value of all initial gifts appears to be small, but the success of the hunt depended on them. All gifts were intended for the owners of the game, who would then give the game to the hunters. Rasmussen (1931, p. 15) also refers to the initial gifts, soles, and wicks that open the caribou season in his discussion of the Nattilingmiut.

Among the Nattilingmiut, the caribou themselves, instead of the mother of the caribou, appear to be the receivers of the gifts. But gifts were not reserved exclusively for either the mother of the caribou or the caribou. Rasmussen (1929) recalled,

> When a caribou killed with an arrow is cut up, care must be taken not to break any of the bones, and when the animal has been cut up, a small piece of meat or suet is placed under a stone – *qingaluklune* – as a sacrifice to the dead – *nErErquvlugit – piʃäkʃalik – manisinialungmat* – it is desired that the dead shall eat, in the hope that they will procure game. (p. 194)

Michel Kanajuq from Qamanittuaq explained how the deceased would interfere on behalf of their descendants:

> Some said that the mother of the caribou lived in heaven, and they thought that she was Nuliajuk. Our ancestors thought that all the people

> who died went to the underworld instead of up above, and that they went to Nuliajuk the mother of the caribou. When the caribou did not want to leave, their faces would change into human-like faces, and the people there would whip them so that they would go to where people were living. An elder would make a small whip, and whip the land, and give it to Nuliajuk and say, 'Here is your whip. Now let us have some caribou'. That is what they used to do to get the caribou to go where they lived. (Laugrand & Oosten, 2009, p. 79)

Kanajuq is our only source for these rituals, however. Apparently, the deceased would use a whip to drive the caribou to the land. The whipping of the land probably prepared the land for the caribou hunting and evokes the use of the whip in rituals to calm the wind. The gift of the whip to Nuliajuk appears to be an initial gift preceding the hunting season.

Thus, relations with the mother of the caribou, the caribou themselves, and the deceased all come into play in the initial gifts required for caribou hunting. Such a configuration indicates that the idea that the animal, and especially the caribou, gives itself to the hunter is probably much more complex. In contrast to Tim Ingold's (2000, p. 13–14) argument, the relationship between the caribou and the hunter cannot be reduced as a discourse or an allegory as many rules were supposed to be followed. Inuit assumed that even the best hunters would be unsuccessful if animals decide to avoid them. But if human beings showed respect to an animal, behaved themselves in accordance with the rules of the ancestors, maintained the important separations between land game and sea game, between everything connected with hunting and all things concerned with human birth, menstruation, and death, the game might allow itself to be killed. A hunter should also show an attitude of humility, imploring the game to come to him and to show itself to him. Finally, the hunter should show great joy at the catch and respect the body of the animal once it was killed. Not killing the animal would imply slighting it. If the game did not feel welcome, it would not show itself again.

Offerings played an important part in caribou hunting. They could be made to the deceased, to the owners of the game, or to the animals themselves. The owners of the caribou demanded offerings of soles and lines before the beginning of the caribou season (Peck, quoted in Laugrand, Oosten & Trudel., 2006, p. 388–89), and *kiversautit*, a small offering was made to the sea woman before the beginning of the seal season. These small offerings were required for successful hunting. As Kappianaq from Iglulik recalled, these small offerings (*tunillai*) were required for successful hunting (Laugrand & Oosten, 2014).

In their article *Sacrifice as the ideal hunt: A cosmological explanation for the origin of reindeer domestication*, Willerslev, Vitebsky & Alekseyev (2014, p. 8) suggest that there is an inherent double bind in the hunt, 'a sort of irreductible paradox', in the fact that on one side, the animal is said to offer itself to the hunter, and on the other hand, the hunter has to draw an arsenal of devious strategies to get its prey. They argue for a structural identity between hunting and sacrifice The description above, however, do not seem to evidence this double bind, nor to this equivalence. On the contrary, among Inuit hunting and 'sacrifice' – I prefer to call it offering – coexist. Moreover, some violence is clearly acknowledged in the killing and not denied. Peter Suvaksiuq from Arviat added that caribou hunts can be scary, as the animals have their own agency:

> What I know is that when there is loud thunder far away when there are no clouds or rain, the caribou are coming towards us. They say that there would be a large herd coming towards us because there was thundering far away ... When there is a herd of bulls together they are scary ... Even though they do not communicate like us, they know what they are doing. (Quoted in Laugrand & Oosten, 2014, p. 211)

Caribou are thus credited with agency; they follow their bulls, their leaders. Inuit hunters acknowledge caribou autonomy.

3 Caribou and Non-social Beings: the Danger of the Hunt

Caribou hunts are usually not dangerous, particularly compared to walrus hunts. Nonetheless, hunters have to be very careful, especially when the caribou are in a big moving herd. We will see that caribou are also connected to the *ijitait*, dangerous mountain spirits. When an animal is seen, the hunter has to make sure it not an *ijiraq*. As with any other hunt, it is impossible to predict the result of the hunt. On certain occasions, however, hunters are informed of how the hunt will go, and having some prediction of the hunt appears to be ideal. Dreams, for instance, predict hunters' fortunes. Angmaalik, an Elder from south Baffin Island, recounted,

> I remember I dreamed of a huge police officer and that day I caught a male caribou with lots of fat. That was my interpretation. If I had a dream of a female non-Native with children, I probably would have seen caribou like that ... If people knew that I had a dream of Non-Natives, the other

> hunters would expect to see caribou that day. (Quoted in Kulchyski et al., 2003, p. 283)

Pisuk from Rankin Inlet recalled,

> If we were out somewhere where there were no caribou, and I dreamed about a little rainbow, then I would know that we were going towards caribou. Sometimes it would be a one-day trip from camp, so we would have to sleep along the way. We would be able to go where there were caribou based on my dream. (Quoted in Kolb & Law, 2001, p. 4)

Other signs might also predict a catch. Kappianaq from Iglulik remembered, 'If my middle finger were to twitch, I used that as a *niriujaarniq* for caribou. If it twitched or was tingling then I knew I was going to get a caribou' (quoted in Kolb & Law, 2001, p. 91). In extreme situations, such as during periods of famine, caribou could also be obtained with *irinaliutiit*, powerful words, or using Christian hymns and prayers, but these options would be only be used as a last resort. Angutinngurniq explained, 'It was not only the *angakuit*; those who were desperate could call the animals as well'. Interestingly, he added,

> I heard this from my mother. She said that her father used to call game. Sometimes in the spring when he became older if he missed a caribou, he had an *irinaliuti* for calling it back. When he went back out, the caribou he missed would be there. But those caribou were never really good. Maybe it was the flesh without the *tarniq* that made it not good to eat. (Quoted in Oosten & Laugrand, 2002, p. 112)

Thus, a caribou without *tarniq*, a soul (or a miniature image of its body), is easier to catch but not good to eat. This statement requires elaboration, but illustrates again this ambiguity of the caribou that is not only flesh. The *tarniq* would give the caribou its capacity to retaliate in case of bad treatment, but also a taste and a connection.

Inuit were aware that caribou sometimes became scarce, but they did not attribute scarcity to an overkill of caribou. Alain Ijiraq (in Laugrand & Oosten, 2014) described how the caribou disappeared after his father used powerful words and the owner of the caribou appeared to him in a dream. When his father was preparing to camp away from the main camp, he heard rumbling sounds from under a group of rocks. The sound became louder until it was deafening. The sound became louder until it was deafening, and sparks shot from the stones. The sparks suggest the presence of *apsait*, nonhuman beings who cause the earth to shake.

Periodic scarcity as well as abundance of the caribou were often attributed to the owner of the caribou, not to overkill. Again, Pinga, the owner of the caribou, would closely watch the hunter's behaviour.

Rasmussen (1929, p. 113) observed that among the Iglulingmiut, that the *ijirait* were among the most powerful helping spirits of the *angakkuit* (see also Boas, 1888/1964, p. 232; 1907, p. 213–214). Rasmussen (1929) glossed the meaning of *ijirait* as 'those who have something about the eyes' (p. 204), explaining that their eyes and mouths were not set transversely in their faces but lengthwise so that they 'blink sideways' (p. 204). *Ijirait* have excellent eyesight. They live inside the hills, which they have fitted up like great stone houses, much resembling those inhabited by white men. They are not visible to human beings – only to shamans – and one should never show fright when one hears them, for they only attack the timid and cowardly.

Ijirait look like humans and live like humans, but they often take on the shape of caribou. Their faces look like the muzzles of caribou, they use caribou skins as clothes, and they dislike everything originating from the sea. Rasmussen (1929) reports, 'The men are dressed in human fashion, only the women's garments are otherwise, their breeches consisting of the white skin from the belly of the caribou; white belly skins all cut up into strips' (p. 205). Thus, the women are dressed with the *pukiq* (white skin strings from the belly of the caribou) that is used in many ritual contexts. According to Rasmussen (1929), 'They are as strong as wolves; when they have killed a caribou, they run home with it, slinging it over their shoulders just as a wolf does with its prey' (p. 205).

George Tataniq from Qamanittuaq also mentioned that *ijirait* live in the ground. He explained that they were known to visit human beings and to lure them back to their houses by giving them a drink (Laugrand & Oosten, 2010, p. 77–78). Iqallijuq associated the *ijirait* recalled how her husband used to live with them. People who meet *ijirait* tend to forget about them. Itinnuaq from Rankin Inlet related that he killed a bull that was probably an *ijiraq*, but he forgot about the details. Taparti observed, 'I have heard that *ijirait* make you forget' (quoted in Oosten & Laugrand, 2002, p. 132–134).

The notion that *ijirait* capture people is a recurrent feature of *ijirait* stories. Pujuat Taparti from Rankin Inlet refers to a human whistling sound as the indication of nearby *ijirait* (quoted in Oosten & Laugrand, 2002, p. 132). When they hear whistling in the air, ordinary people are afraid of that *ijirait* are close by. *Ijirait* are fast runners and can overtake caribou easily. When they capture human beings, they make them into fast runners by having the flesh from their shinbones or toes eaten away by worms in the earth or tiny creatures in the lakes (Rasmussen, 1929, p. 205). This method of transforming a human being into an *ijiraq* evokes techniques of shamanic initiations where the flesh of arms or hands is eaten by insects or worms.

In an account of a meeting with *ijirait*, the *angakkuq* Oqamineq related, 'They had a small pocket in their tunics, in which they kept two small stones. They were mighty runners, and could outrun the caribou; when they came up quite close, they killed them with the stones in their pockets' (Rasmussen, 1929, p. 206). Rasmussen (1929) also reported,

> Shamans have found among them strange implements very much like the mirrors used by whitemen. This implement is at once a mirror and a spyglass. It glitters like mica, and when one looks down into it, all that is passing far away among the dwellings of men is reflected in this mirror; therefore the ijerqat know all about mankind. (p. 204–205)

Thus, unlike caribou, the *ijirait* have extraordinary eyesight.

Deceased people may join the *ijirait* (see Saladin d'Anglure, 1983). Nutaraaluk from Iqaluit suggested, 'Maybe they are the deceased's people's *tarniit* that remain on the land' (quoted in Saladin d'Anglure, 2001, p. 54). And Aupilaarjuk from Rankin Inlet stated, 'Dead people turn into *ijirait*. That is what I have heard. I have not seen this personally, but I believe it ... [*I*]*jirait* are said to be deceased people that turn into caribou' (quoted in Saladin d'Anglure, 2001, p. 56, 60). Nutaraaluk suggested that *ijirait* live forever: 'Some of those who roam the earth are said to never leave it' (quoted in Saladin d'Anglure, 2001, p. 55). Aupilaarjuk explained,

> They can go to the store, but no matter how many times they shop, the stock never depletes. Over time they can buy snowmobiles just like we do. When we had dog teams, they had dog teams. When we no longer used dog teams and we used snowmobiles, they used snowmobiles. (Quoted in Saladin d'Anglure, 2001, p. 55).

Thus, Inuit correlate *ijirait* development with that of ordinary human beings.

Encounters with *ijirait* are numerous in Inuit oral traditions and violence is often involved. Rasmussen (1929, p. 205–206) recalled an encounter of the *angakkuq* Qingailisaq or Oqaminiq, the father of the *angakkuq* Aava, Rasmussen's main informant, with the *ijirait*. One *ijiraq* tried to throw him down, but the *angakkuq* withstood the attack. They separated on friendly terms, and the *angakkuq* had a coat made with ornamentations that depicted the encounter.

Today, Elders still have many experiences with *ijirait*. Angutinngurniq from Kugaaruk related,

> Because my mother was named Ijiraq, I often thought that maybe it was because of her that I saw *ijirait*. I have seen them more than once. Even though they are close by, and there isn't anywhere for them to hide, they tend to disappear. (Quoted in Oosten & Laugrand, 2007, p. 167).

John Makitgaq from Qamanittuaq remembered he had seen one and 'its legs were like the kamiks that women make from caribou leg skins.' (quoted in Mannik, 1998, p. 139). Kappianaq gave an extensive account of an encounter with *ijirait*: 'An *ijiraq* is able to make the land ripple … I got lost for three days because of them … I know that there are *ijirait* even though I am Christian' (quoted in Oosten & Laugrand, 2001, p. 71–74). Aupilaarjuk commented that although *ijirait* can easily retaliate if one of them is killed, Inuit who had a tattoo between the eyes, a *kigjugaq*, were usually protected from their attacks. The marking provided protection (quoted in Saladin d'Anglure, 2001, p. 60, 62).

The *ijirait* thus appear as caribou people; they share features of caribou but, like humans, they also hunt them. Furthermore, human beings can join the *ijirait* after death. The contrast between *ijirait* and animals is marked by eyesight; while caribou see very poorly, *ijirait* have excellent eyesight.

Although humans and caribou might seem to have a straightforward hunter–prey relation, in fact, this relation is complicated by the *ijirait*, who appear to identify both as caribou, in their capacity to transform into caribou, and as deceased human beings, in the fact that deceased as well as living human beings can join them. When they transform into caribou, they should not be killed, and they are very capable caribou hunters themselves.

4 Following and Respecting the Caribou

In Siberia, hunting and pastoralism are not mutually exclusive (Stépanoff et al., 2017, p. 64), but in the Canadian eastern Arctic, Inuit were never attracted by the idea of becoming reindeer herders. Like the Yup'ik, they always maintained that they preferred to hunt caribou rather than herd it. In an interview discussing the Amadjuak experiment, when Saami were brought to Baffin Island, the hunter Pauloosie Kilabuk stated,

> *Nuvukiulaq* [the Saami people, literally 'that resembles to points' because of their four points hats and their boots] were friendly people. Inuit made friends with them and enjoyed staying in their camp. People in that area found their clothing amazing, especially the way they make them.

> Saami people were different than the Inuit. They lived with similar animals but in a different way. I think the only difference is that you cannot get close to a caribou the way you can get close of a reindeer. Usually the reindeer has also more fat than the ground caribou you get here. I heard that there were some Inuit who brought the Saami into the land, so some Inuit were involved in helping the Saami. An elder from Cape Dorset could remember about Simiuni Qupaapik who died long time ago and was one of these people who worked with the Saami. Some Saami could learn Inuktitut because they lived with Inuit. We never talked English in those days. That's the way it happened. When the Saami left, the Inuit took care of the reindeer for a while and enjoyed the work. But I think, we Inuit want to kill animals, not to live with them. That's the only way. We are alive today thanks to the animals. Back then, there was probably nobody to look after the reindeer so Inuit left them. (Interview, March 31, 2006)

Although Inuit felt closely related to the Saami, the idea of herding clearly did not appeal to them. Later in the interview, Kilabuk added that caribou meat was much better than reindeer meat: 'Some of the reindeer were killed to be eaten. The meat doesn't have as much flavour as the caribou we have here. It's a little bit watery. Our caribou is absolutely very much better' (Interview, March 31, 2006). According to Kilabuk, caribou and reindeer later interbred:

> Something I heard about this story is that a caribou around here was found from those reindeer. When the Sami left, the reindeer started mixing with the caribou. Now, there are some caribou mixed from reindeer. We see them sometimes. We know those are mixed but we would hunt them as well, even if the caribou meat is much better. We know that, we have been eating caribou for many years. They were here first and we prefer eating them. (Interview, March 31, 2006)

Kilabuk also insisted that Inuit never like the idea of seeing the animal mating; it felt too intrusive:

> At that time they were trying to increase the caribou here so they got the caribou and the reindeer together. The animals seemed kind of embarrassed by the people watching as they were getting together. They were looking at the people and were really embarrassed. (Interview, March 31, 2006)

Having to closely watch the animals and keep them in the vicinity of their camps were never considered fruitful activities. On the contrary, Inuit enjoyed their nomadic life and following the animals. They respected their perceptions and desires, their agency.

This does not mean, however, that Inuit ignored domestication (after all, they used dogs) or refused to adopt animals. Adopting an animal was possible for short periods. For example, Inuit adopted bear cubs, wolves, or even falcons to better hunt animals. They also sometimes adopted caribou, as Simon Tookoome (1999, p. 24), an Elder from Qamanittuaq, mentioned. While adoption was possible at an individual level or in exceptional circumstances, it was never considered a collective option. Inuit always preferred to be hunters, not herders, even if they had to face starvation, as was the case in the early 1920s.

At a cosmological level, the transition from hunting to pastoralism is very complex, and this article only introduces some context for understanding this shift. But the perspectives brought by Ingold and by Willerslev, Vitebsky and Alekseyev do not help to explain how Inuit relate to the caribou. With respect to Descola's argument, Inuit are closer to the Chukchi than to the Innu or the Exirit-Bulagat models (see also Descola, 2005, p. 513–514), yet they resented domestication.

5 Conclusion

Why Inuit hunters resist herding so much is difficult to explain. One could argue that they are in fact already familiar with the idea of mastering animals, using them for transport, as illustrated by their use of dogs. Using Taber's paper, Ingold (2000, p. 73), who studied the transition from hunting to pastoralism in North America and Eurasia, suggests that domestication implies 'a kind of mastery and control similar to that entailed in slavery' (p. 73). Inuit never developed that dominant relation of production, and they were probably always very reluctant to such a system. They preferred to remain predators. Similarly, Inuit never saw themselves as people who protect the caribou, as this scheme of protection seems essential in herding cultures (Ingold, 1980, p. 27; Descola, 2005). Inuit are more inclined to think that caribou have their own mind, they are subjects and that humans can only follow them and respect their *nabluit*, their paths. They might be connected to an owner but this connection is not straightforward. We do not find any double bind as Willerslev, Vitebsky and Alekseyev (2014) characterize it. Inuit consider caribou autonomous; they cannot be appropriated as they remain closely connected to the land and the deceased. Reindeer herders are seen like 'cousins', yet Inuit identify themselves as hunters. In the past, Inuit performed rituals before the hunt as if they would prefigure the future result of the hunt (see Laugrand & Oosten, 2014). Many cultural and spiritual rules apply to the caribou, and Inuit still respect these rules.

The connection between the caribou and the land is always strong, and *ijirait* seem to be present everywhere in the Canadian Arctic. The *ijirait* are connected to the land, the caribou, and the deceased. They carry connotations of caribou as well as of caribou hunters and may, to some extent, represent the owners of the caribou. But Pukimna or Pinga may also have been owners of the caribou. The deceased can also figure as owners of the caribou, as Kanajuq explained, when they send the caribou to the hunters to be killed. A fixed model of ownership did not exist, and more than one agent could represent the owner(s) of the caribou.

With respect to the Inuit in Canada, not only the *ijirait* but the caribou themselves may also have been connected with the deceased. According to one of the origin myths, the caribou came out of the earth, which also receives the dead bodies of human beings, and caribou bodies were brought in through the back of the igloo, just as the corpses of people, wrapped in a caribou skin, were brought out. The human faces of the caribou described by Kanajuq would also fit into such a pattern, as do Davidialuk Alasuaq's drawings, collected by Bernard Saladin d'Anglure, which inspired Ingold (2000, p. 117–123). The caribou may have represented anonymous deceased ancestors. This interpretation gives more meaning to a painting by Nick Sikkuark (1976) that appears in his booklet *What Animals Think*; the painting depicts a caribou buried with stones, as was usually done for dead humans, with the caption, 'When I was alive, I waited for man to kill me. Now that I am dead, I am still waiting' (Sikkuark, 1976, p. 16).

To conclude, the symbolism of the caribou remains complex. Its *tarniq*, soul, makes it more difficult to catch but also gives more taste to its flesh. According to Inuit hunters, reindeer, by constrast, do not taste good. A strong connection existed between human beings and caribou, shaped by the *pukiq*, the white belly skin of the caribou, which took a central place in shamanism and various rituals. If a *pukiq*, that also designates a white caribou, was killed, mourning rituals evoked a kinship relationship with the deceased animal – specifically the *pukiq* would be treated as a deceased sister. When a *silaaq*, an earth egg, was killed, relations with a deceased brother were evoked by the mourning rules that had to be observed. Thus, in the context of death, these exceptional animals were viewed as close kin. Caribou provided human beings with a second skin that allowed them to survive in the Arctic climate. The caribou clothes could come to life in the context of shamanism and provided the *angakkuit* with tools for communication or other purposes. The belt, made of *pukiq*, and the coat are especially marked in this respect.

Hunting caribou was – and to a large extent still is – considered the only way to preserve them, and Inuit hunters very much dislike the way biologists pursue and capture caribou with helicopters and nets to study their behaviour. The size of caribou herds has been an issue of debate for a long time, and Inuit hunters often disagree with qallunaat (white people) and biologists on caribou population estimates. In Inuit's view, the caribou and their lifestyle must be respected.

Closely associated with *nuna* (the earth), caribou are referred to as the big lice of the earth. Like the *qupirruit*, small beings, they are considered masters of life and death. They always return in great numbers despite the recurrent cycles of increase and decrease that characterize the great herds in the North. They are often defined as a multitude, and the sound of the herds is associated with thunder. Thunder is also connected with the *apsait*, and that connection should be explored further. In the past, complex feasts connecting successful hunting with women's fertility were celebrated. Ritual injunctions ensured that caribou and sea game were separated. Women, especially menstruating and pregnant women, were subject to all kinds of restrictions with respect to the caribou, and these applied particularly to the *nablut*, the crossing places of the caribou, where they used to be killed in multitudes. Today, many of these practices have disappeared, but caribou remain a major prey and remain closely associated with the *ijirait*. Killing them implies some violence – that of a killing – but is also a matter of respect – it avoids depletion of the game – and essential for humans to survive. Yet, from the hunter's perspective, hunting is not sacrifice To what extent the owner of master dominates his herd, as Ingold (2015, p. 24) explores in the Siberian context, cannot be applied here either because caribou are not always associated with an owner. Rather, they are connected to the land and to a multitude. The master is by no means a 'spiritual pastoralist'. On the contrary, and like the Inuit themselves, the master cannot control these animals. The situation is different with seals and sea mammals, but that is another subject.

References

Anderson, D. G. (2000). *Identity and ecology in Arctic Siberia: The number one reindeer brigade*. Oxford University Press.

Birket-Smith, K. (1929). *Report of the Fifth Thule Expedition, 1921–24: Vol. 5, pt. 1 & 2. The Caribou Eskimos: Material and social life and their cultural position* [Descriptive part]. Translated from the Danish by W. E. Calvert. Gyldendalske Boghandel.

Boas, F. (1964). *The central Eskimo*. University of Nebraska Press. (Original work published in 1888).

Boas, F. (1907). The Eskimo of Baffin Land and Hudson Bay, New York. *Bulletin of the American Museum of Natural History, 15*(2), 374–570.

Conaty, G.T., & Lloyd, G. B. (2004). *The reindeer herders of the Mackenzie Delta*. Firefly Books.

Descola, P. (2005). *Par-delà nature et culture*. Gallimard.

Harvey, G. (2017). *Animism: Respecting the living world*. Hurst.

Harvey, G. (2019). Animism and ecology: Participating in the world community. *Ecological Citizen, 3*(1), 79–84. www.ecologicalcitizen.net/article.php?t=animism-ecology-participating-world-community.

Ingold, T. (1980). *Hunters, pastoralists and ranchers: Reindeer economies and their transformations*. Cambridge University Press.

Ingold, T. (2000). *The perception of the environment: Essays in livelihood, dwelling and skill*. Routledge.

Ingold, T. (2015). From the master's point of view: Hunting is sacrifice *Journal of the Royal Anthropological Institute, 21*(1), 24–27. https://doi.org/10.1111/1467-9655.12145

Kolb, S., & Law, S. (Eds.). (2001). *Inuit perspectives on the 20th century: Vol. 3. Dreams and dream interpretation*. Nunavut Arctic College.

Kulchyski, P., McCaskill, D. & Newhouse, D. (Eds.). (2003). *In the words of Elders: Aboriginal cultures in transition*. University of Toronto Press.

Laugrand, F., & Oosten, J. (2008). *The sea woman: Sedna in Inuit shamanism and art in the eastern Arctic*. Alaska University Press.

Laugrand, F., & Oosten, J. (2010). *Inuit shamanism and Christianity: Transitions and transformations*. McGill-Queen's University Press.

Laugrand, F., & Oosten, J. (2014). *Hunters, predators and prey: Inuit perceptions of animals*. Berghahn Books.

Laugrand, F., & Oosten, J. (2015). Les rennes d'Amadjuak. Eleveurs saamis et chasseurs inuits en Terre de Baffin (1921–1925). *Recherches amérindiennes au Québec, 47*(1), 59–74. https://doi.org/10.7202/1035165ar

Laugrand, F., & Oosten, J. (Eds.). (2009). *The ethnographic recordings of Inuit oral traditions by Father Guy Mary-Rousselière*. Nunavut Arctic College.

Laugrand, F., Oosten, J. & Trudel, F. (Eds.). (2006). *Apostle to the Inuit: The journal and ethnographic notes of E.J. Peck: The Baffin years (1894–1905)*. University of Toronto Press.

Low, A.L. (1906). *Cruise of the Neptune, 1903–1904: Report on the Dominion Government Expedition to Hudson Bay and the Arctic Islands on Board the D.G.S. Neptune*. Government Printing Bureau.

Lyon, G.F. (1824). *The private journal of Captain G.F. Lyon of H.M.S. Hecla, during the recent voyage of discovery under Captain Parry*. John Murray.

Mannik, H. (1998). *Inuit Nunamiut: Inland Inuit.* Friesens Corporation.

Oosten, J., & Laugrand, F. (Eds.). (2001). *Inuit perspectives on the 20th century: Vol. 2. Travelling and surviving on our land.* Nunavut Arctic College.

Oosten, J., & Laugrand, F. (Eds.). (2007). *Surviving in different worlds: Transferring Inuit traditions from Elders to youths.* Nunavut Arctic College.

Oosten, J., & Laugrand, F. (Eds.). (2010). *Hardships of the past: recollections of Arviat Elders.* Nunavut Arctic College.

Oosten, J., Laugrand, F. & Rasing, W. (Eds.). (1999). *Perspectives on Traditional Law,* Interviewing Inuit Elders 2. Nunavut Arctic College.

Plattet, P., & Lincoln, A. (Eds.). (2014). Reindeer herding on the Alaska Peninsula [Special section]. *Alaska Journal of Anthropology, 12*(2), 1–73.

Rasmussen, K. (1929). *Report of the Fifth Thule Expedition, 1921–24: Vol. 7, pt. 1. Intellectual culture of the Iglulik Eskimos.* Gyldendalske Boghandel.

Rasmussen, K. (1931). *Report of the Fifth Thule Expedition, 1921–24, Vol. 8, pts. 1–2. The Netsilik Eskimos: Social life and spiritual culture.* Gyldendalske Boghandel.

Saladin d'Anglure, B. (1983). Ijjiqat: Voyage au pays de l'invisible inuit. Études *Inuit Studies, 7*(1), 67–83.

Saladin d'Anglure, B. (Ed.). (2001). *Interviewing Inuit Elders: Vol. 4. Cosmology and shamanism.* Nunavut Arctic College.

Sikkuark, N. (1976). *What animals think: Pictures and stories.* Government of NWT, Department of Education.

Stépanoff, C., Marchina, C., Fossier, C., & Bureau, N. (2017). Animal autonomy and intermittent coexistences: North Asian modes of herding. *Current Anthropology, 58*(1), 57–70. https://doi.org/10.1086/690120

Tookoome, S. (1999). *The shaman's nephew: A life in the Far North.* Stoddart Kids.

Tuck, J. (1954). *The Baffin Island reindeer experiment, 1954.* Ms. MSS-158, Dartmouth College Library, Hanover, NH.

Vaté, V. (2007). The Kêly and the fire: An attempt at approaching Chukchi representations of spirits. In F. Laugrand and J. Oosten (Eds.), *La nature des esprits dans les cosmologies autochtones* (pp. 219–238). Presses de l'Université Laval.

Vitebsky, P. (2005). *The Reindeer People. Living with animals and spirits in Siberia.* Houchton Mifflin Company.

Willerslev, R., P. Vitebsky, & Alekseyev, A. (2014, Nov. 19). Sacrifice as the ideal hunt: A cosmological explanation for the origin of reindeer domestication. *Journal of the Royal Anthropological Institute, 21*(1):1–23. https://doi.org/10.1111/1467-9655.12142

Willerslev, R., P. Vitebsky, & Alekseyev, A. (2015, Jan. 29). Defending the thesis on the 'hunter's double bind'. *Journal of the Royal Anthropological Institute, 21*(1): 28–31. https://doi.org/10.1111/1467-9655.12146

CHAPTER 4

Too Many *Onças*: Taxonomical Dilemmas among the Karitiana in Southwestern Brazilian Amazon

Felipe Vander Velden

Abstract

The Karitiana, an Amerindian people in the southwestern Brazilian Amazon, mention the existence of many types of *onças* (jaguars) with whom they inhabit the world. While many are not recognized by scientific or popular/regional zoology, they are part of the world experienced by the Karitiana (and other Amazonian peoples). This article suggests that these Karitiana *onças* should be understood through the intersection between the experience and detailed knowledge these Indigenous people have of the territory in which they live, and the enormous richness and complexity of the world and the forms of life that exist. Rather than simply arguing that the Karitiana know a world that science cannot know (because of ontological and/or epistemological differences), the suggestion is that the world is so rich and has so many facets that there are countless ways of observing and understanding it.

Keywords

jaguars – taxonomy – ontology – Karitiana – Amazonia – knowledge

•••

> There are greater differences between a plough horse or draft horse and a racehorse than between an ox and a plough horse.
>
> GILLES DELEUZE, *Spinoza: Practical Philosophy*

∴

 | DOI:10.1163/9789004679450_005

1 Introduction

There is a surprising number of *onças* in Indigenous Amazonia. The anthropological literature written in Portuguese refers to *jaguares* rather than *onças* (where both words mean jaguars), since most Brazilian ethnologists use the former term (a masculine noun originally from the Tupi *ya'wara*, which is used in European Portuguese and standard English) rather than the latter, which is more common in Brazilian Portuguese (a feminine noun, from the Greek *lyqx* and the source of the Latin *luncea* and Italian *lonza*).[1] Although there are many jaguars and few *onças* in the ethnographic literature (and, unfortunately, fewer and fewer of the animals themselves in our neotropical ecosystems), the Karitiana present an interesting alternative on this topic. The Karitiana (*Yjxa*) are a Tupi-Arikém-speaking Indigenous People who live in seven villages outside the capital of the state of Rondônia, Porto Velho, in the southwestern Brazilian Amazon (see Figure 4.1); today they number approximately 450. In their experience of the world, there is, in fact, an excess of *onças* (at least from a scientific zoological point of view); the Karitiana acknowledge not just one 'type', but several in their own language, which as we shall see can be difficult to translate. The Karitiana use the term *tipo* (type) in Portuguese to designate the existence of certain beings that are recognisably different from others. It is generally used in expressions such as '*o gato-maracajá é* [um] *tipo de onça*' ('a margay is a type of a jaguar'). These analogies are sometimes also signalled with the expression '*X é parente*' ('X is kin'): '*o gato-maracajá é parente da onça*' ('the margay is kin to the jaguar').

This article discusses the apparent, unexpected proliferation of these beings in the world of the Karitiana at a sort of frontier between the real and the fantastic, between the regular and the monstrous, at a point where the world surprises us with its rich, multifaceted, unexpected possibilities. The hypothesis is that the attention paid by the Karitiana (like other Indigenous Peoples around the world) to diversity, and the variety of nonhuman animal forms that inhabit the world, leads them to divide the world differently, establishing within it discontinuities which are different from those recognized by both scientific taxonomy as well as from other non-Indigenous classification

1 This is in line with the *Novo Dicionário de Língua Portuguesa* by Aurélio Buarque de Holanda. Note that, in early usage, the Tupi term came to designate the introduced domestic dog (a case of 'marking reversal' *sensu* Brown, 1999), resulting in the jaguar being designated *jawara -ete* or *jaguaretê*, as is still heard in certain parts of Brazil (for example, in the 1962 story by Guimarães Rosa 'Meu tio, o Iauaretê'), which reflects the prevailing use in the centre-north region of Minas Gerais.

systems. One example of such systems is Brazil's regional folk-taxonomy, which uses a kind of hybrid language (including Indigenous as well as Portuguese terms, with many animal names derived from native languages) that is also used by the bilingual Karitiana, and as we shall see, is a product of the encounter between Indigenous and European knowledge in colonial Brazil. The objective is to suggest that the world and its creatures comprise such a staggering variety and diversity of forms and agencies that directly impact the perceptions human groups have of them. Combining Amerindian ethnology and anthropological studies on the relations between human and nonhuman animals can make significant contributions to both fields, which for the most part remain separate.

Taxonomically speaking, *onças* (*Panthera onca*; Linnaeus, 1758) are carnivorous mammals of the family Felidae; they are the only large cats in the genus *Panthera* native to the western hemisphere (Johnson, 2016, p. 7). They are the largest and most powerful predators in the Americas, where they ranged extensively in the past (from the southern United States to the southern central region of Argentina and Chile, and throughout all of what is now Brazil). However, their numbers have been decimated by indiscriminate hunting for their pelts (which despite bans still garner high prices in the luxury fur market), out of fear, or to protect domestic animals and, ultimately, people. Note that the term *onça* (in this text and in general discourse) refers to an animal which is spotted, even though Brazilian vernacular speech uses the same term to refer to dun-coloured cats of other species – *Puma concolor* (the cougar), known in Portuguese as *puma* or *suçuarana*, and other smaller native felines: the ocelot (*Leopardus pardalis*), the margay (*Leopardus wiedii*, see Figure 4.2), Geoffroy's cat (*Leopardus geoffroyi*), the oncilla (*Leopardus tigrinus*), the jaguarundi (*Puma yagouaroundi*), and the pampas cat (*Leopardus colocolo*) (Emmons & Feer, 1997). The first four are frequently designated in Brazil as *onças* (or *onçinhas*, 'little jaguars') since their markings resemble those of the spotted *onças*.[2] In this article, the term specifically refers to *Panthera onca*, a beautiful and ferocious nonhuman animal, a charismatic predator with enormous cultural importance to Indigenous and non-Indigenous rural people in various parts of Brazil, particularly the Pantanal wetlands of Mato Grosso and Amazonia (Gambarini, Duarte, Haberfeld, & Cunha de Paula, 2016).

2 While there is no space to illustrate all the animals mentioned in this text, they can be easily found on the internet or in books and field guides such as Emmons & Feer (1997) and Einseberg & Redford (1989, 1992, 1999).

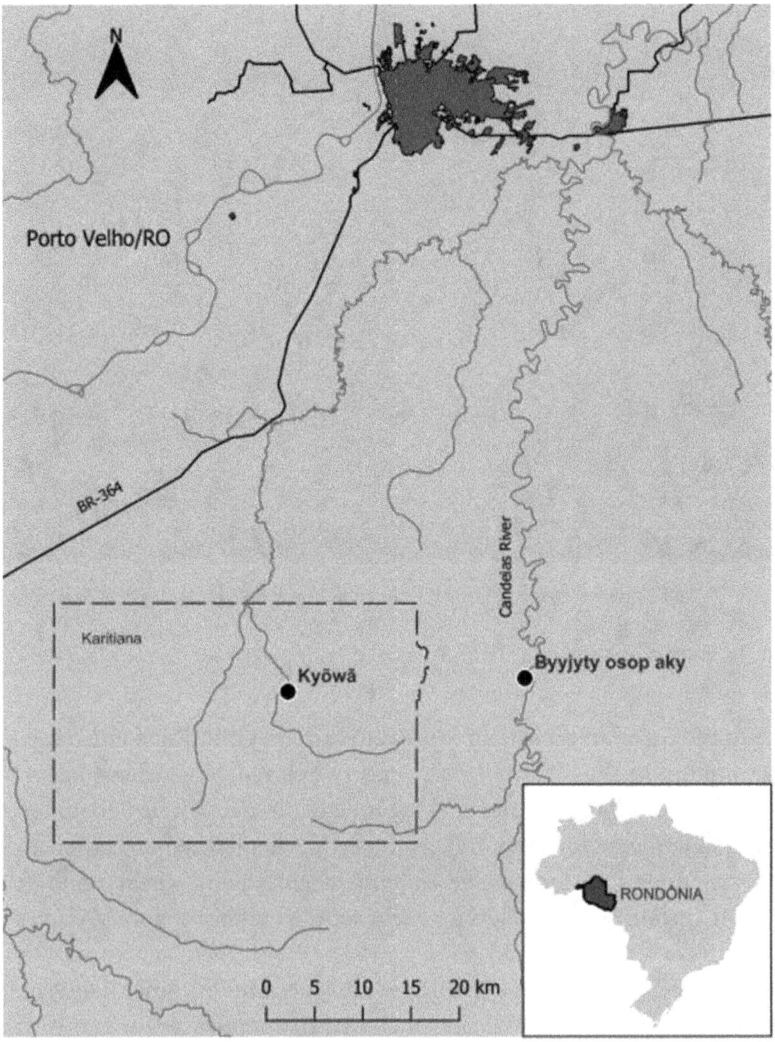

FIGURE 4.1 Map showing the location of the Karitiana Indigenous Land and the two villages where fieldwork was carried
SOURCE: MAP BY WELLINTON COSTA SILVA

The history of the terms is interesting; they can be considered cultural artifacts, since 'species' is a concept stemming from the scientific worldview (the notion of species has consequently come under critical scrutiny from contemporary zoology itself, cf. Pavlinov, 2013; Ritvo, 2018). The etymology of the word 'jaguar' in English, with its obvious Tupi-Guaraní origin, suggests that the taxonomic category was incorporated into European zoological knowledge from South America, from speakers of these languages. This may have occurred via

FIGURE 4.2 Two Karitiana boys exhibiting a head of a margay they have just hunted, *Kyõwã* village, 2003
PHOTO: F. VANDER VELDEN, 2003

Ulrich Schmidl, a German soldier known to have been in the southern part of the continent in the mid-sixteenth century. Schmidl would have heard of these animals from the Guaraní-speaking Indigenous Peoples he encountered in the Bacia Platina region (Schmidl, 2008): 'In a general way, one can say that Guaraní denominations form a well elaborated system and – *cum grano salis* – display a certain resemblance to our scientific nomenclature.' (Dennler, 1939, cited in Lévi-Strauss, 2008 [1962], p. 61). Later, the Swedish naturalist Carl von Linnaeus incorporated the taxonomic category into his binomial system, initially as *Felis onca*, which was later modified to *Panthera onca*. In this way, a zoological species which was already intimately associated with an Indigenous 'species' or 'type' came into being.[3] A portion of the world identified by the Guaraní – a creature designated *yaguar* or *yaguarete* – was borrowed and obviously transformed during the process, by a German explorer and then

3 'The etymology of the word *jaguar* is unclear. Some sources suggest a borrowing from the South American Tupi language to English via Portuguese, while others attribute the term to the related Guarani languages. In the Tupi language, the original and complete indigenous name for the species is *jaguar*, which has been reported as a denotation for any carnivorous animal – in the compound form *jaguareté*, – *eté* means 'true'. In the related Guaraní languages, *yaguareté* has been variously translated as 'the really fierce beast', 'dog-bodied', or 'fierce dog'' (The New World Encyclopedia, available at http://www.newworldencyclopedia.org/entry/jaguar#Etymology, retrieved September 24, 2018, italics in original).

a Swedish naturalist, to become another name in a hierarchically organized collection (a taxonomy) within the still-nascent science of modern zoology. Nevertheless, when taken out of its South American ecological and social context and subjected by the totalising language of European natural history (Pratt, 1999), this designation becomes fundamentally altered; it leaves the world to enter the kingdom of words, the pure signs of seventeenth-century science (Foucault, 1994). The native terms (*onça, jaguar*) are nearly encapsulated in the taxon *Panthera onca*—a dialog between local and scientific forms that was common throughout the history of taxonomy (Atran & Medin, 2008, p. 25–28)—although never completely, as we shall see. To paraphrase Maurer (2000, p. 691), and taking for granted an assumed realism which I am unable to discuss adequately here (but which is described by Kohn, 2013), *onças* are *onças* and only certain very specific relations transform them into *Panthera onca* (or into jaguars, *yaguar*, *yaguarete*, or *obaky*).

Here we do not address the specific process which most likely compressed and standardized a group of understandings held among southern Guaraní peoples related to the *onça* (or *yawara*) and other possible correlates into the scientific taxon *Panthera onca*. It is interesting to note that there are also several types of *onça* among the Guaraní (Lévi-Strauss, 2004, p. 83), and that the names of the various species (as they are formally recognized today) appear to have circulated widely across the different continents: just as the term *onça* was used by the Portuguese for the large cats of Africa and Asia (leopard or panther, cheetah, lynxes, caracal), *tigre* (tiger) was used for the jaguars and pumas in southern Brazil and the Andes (Papavero, 2015, 2017). But our main question here is specific to one group and one term: what is this animal or life form we call *onça* to the Karitiana (who use this term rather than *jaguar* when they speak Portuguese), and how did it come to be? Or, in other words, what is at play when the Karitiana discuss this predator which is both feared and respected, materially, and symbolically important in terms of indigenous, popular, and scientific knowledge?

2 Materials and Methods

The data discussed here were collected during various periods of research conducted between 2003 and 2015 in two of the Karitiana villages, *Kyõwã* and *Byyjyty osop aky*, in the north of the state of Rondônia, in the Brazilian Amazon, totalling 13 months of field work. The classic ethnographic method was used, with participant observation, semi-structured interviews (mostly with male hunters spanning 20 to 60 years of age, all bilingual in Karitiana and Portuguese), and long conversations with one or more people, as well as the use of field guides to identify and discuss nonhuman animal species (Emmons & Feer, 1997, for neotropical mammals).

3 *Onças*, Jaguars and Distinct Types of *Obaky*

When they use the term *onça* in Portuguese, the Karitiana refer to spotted *onças* by default: these are designated by the term *obaky* in their Indigenous language. These animals (sometimes also called *obaky pe'ejema*, '*onça* with a spotted design') are greatly feared, and Karitiana hunters always keep an extra cartridge in their shotguns after a hunt in case of a chance encounter. They say the meat is not suitable for eating ('*não presta*'), which is the case with most carnivorous species throughout the world (Foer, 2010). In general, it is said that *onças* are not *himo* (a Karitiana word designating both 'hunting prey' and 'meat'), although some concede they could be prey but are '*outro tipo*' (another kind), indicating their non-edibility (*him sara*, 'bad meat' or *hima a'oky* 'inedible meat'). My data suggest that *onça* meat is inappropriate for eating for two reasons: first, it causes people to develop a respiratory problem and tire very quickly (*sombyko*, 'without breath') as they go about their activities (because like domestic dogs, *onças* always have their mouths open, panting). Additionally, eating this meat can have fatal consequences because *onças* (also like dogs, the Karitiana say) do not care for their offspring, who 'often die' (here they are contrasted with tayras [*Eira barbara*], which protect their young and consequently can be eaten by the Karitiana). Anyone who eats jaguar meat gets a 'large angry, open wound, a tumour' (*patirity*). Perhaps because *onças* are 'chiefs' (that is, the 'owners' of prey animals, as described below), their 'blood is different', and their meat consequently provokes wounds.

Onças are considered to have a strong smell, like rotten fish (*opira*), which may be related to the many ambiguous associations between *onças* and the dead: *onça* teeth are placed in the mouths of the dead, both men and women, and there are indications that one of the souls of the dead (the Karitiana state that all persons have four 'souls' or 'shadows') leaves the corpse violently and transforms into a huge, very aggressive and cannibalistic *onça* that patrols the perimeter of villages on the silent nights that follow a death. These same feline teeth are highly valued by men, since necklaces made from them are the ideal present for a man to give to his father-in-law; because of these teeth (above all the canines), even small *onças* are always killed whenever they are encountered. The Karitiana say that the necklaces of jaguar teeth used in the *chicha* festivals (rituals in which large quantities of manioc beer are consumed to the sound of long bamboo flutes) also drink *chicha* and should be fed the blood of prey animals from time to time, for reasons that are unclear.

Nevertheless, many Karitiana say that the most dangerous animals along the forest trails are snakes, above all the *picos-de-jaca* (the popular name for both *Lachesis muta* and *Bothrops atrox*) hidden among the trees; in contrast, they say that *onças* are easily seen in the forest, and when 'carrying weapons [especially firearms] one does not fear *onças*'. What sets *onças* apart is the ferocity of their behaviour – bloodthirsty hunting, eating alone, attacking furtively, or 'seeing without being seen' (Süssekind, 2014) – which makes them the complete opposite of sociable, compassionate, and (ideally) peaceful human beings. These Karitiana notions correspond with how jaguars are considered in many other Amerindian cosmologies from all over South America (Saunders, 1998; Londoño Sulkin, 2005; Santos-Granero & Mentore, 2006; Fausto, 2014, among others).

To a certain extent, the beings recognized as *onças* (including those seen as similar or 'related') by the Karitiana can be individually correlated with zoologically distinct species – which the Karitiana themselves do – although some may not be considered true *onças*. This is the case with the weasel-like tayra (*Eira barbara*), which is technically a mustelid but called the 'black *onça*' (*obaky emo*) by the Karitiana; the crab-eating racoon, a procyonid carnivore (*Procyon cancrivorus*) they call '*onça* coati' (*obaky iri'sa*); and two animals they call 'wild dog', 'fox', or 'wolf that is like a dog', in Portuguese and *kypõn* and *gyryty* in their own language, probably native canine species considered to be 'like *onças*'. Domestic dogs are also considered to be like onças, as has been recorded among other Amerindian peoples such as the Achuar (Descola, 1994, pp. 84–86), the Guaraní (Lévi-Strauss, 2004, p. 83), and the Chané in the Chaco (Villar, 2005). Dogs are *onças*, but are special and different from their counterparts in the forest (*gopit*, 'lives in the bushes'): they are termed *obaky by'edna*, '*onças de criação*', literally translated as, 'nurtured *onças*', 'domestic *onças*', or according to the Karitiana, 'tame *onças*'. This indicates an intrinsic ambiguity of these animals, companions of human beings, which are both sociable and ferocious, docile and predatory (Vander Velden, 2017).

Cougars or 'dun-coloured *onças*' (*obaky somo,* literally 'vermilion *onças*'), feline margays (*obaky ina*, '*onçinha*', 'little *onça*'), as well as domestic cats (*Felis catus*) introduced by the whites (which are also called *obaky ina* but do not appear to be confused with margays by the Karitiana, despite the same name) are also *onças*, as are the 'black *onças*' (*torowoto* or *obaky torowot*, that also should not be confused with tayras, which the Karitiana also call 'onças-pretas' in Portuguese). They also speak of 'large black *onças*' (*obaky ty'emo*), although here the Karitiana classification differs from the zoological taxonomy: to them black *onças* are different beings than spotted ones, while modern zoology considers both to be the same species with varying concentrations of melanin in their coats.

The intention here is not to definitively equate 'species' of Amazonian fauna with the 'species' or 'types' known to the Karitiana (Vander Velden, 2012); in fact, the quotation marks here indicate the precariousness of the 'species' concept in contemporary discussions. Here this type of equivalence between native types and zoological species is eschewed to avoid deeming the 'species' real or natural and discounting the reality of indigenous perceptions (understanding them only as cultural variations of a scientific reality); even zoological taxonomy recognizes the eminently transitory nature of distinct species (Ritvo, 2018). The Karitiana constantly alter the system, causing a proliferation of these beings, since in their careful attention to the variations inherent to the fluctuations of life and ecologies (as we shall see) they do not seem to establish or consolidate taxa by isolating them from their multiple relationships and potential affections.

4 Animals, Monsters, '*Bichos*:' a Karitiana Bestiary

Indeed, there are many more types of *onças* in the Karitiana world. They describe this assemblage of *onças* as '*tudo bicho*'; *bicho* is a general term in Portuguese that can mean 'beast' or 'bug', but here is used in a different sense to indicate a monstrous being. All these beings are greatly feared for their unpredictable ferocity and have inedible meat: *obaky bijo* ('smelly-monkey-*onça*'), *obaky oromo* ('monkey-*onça*'), *obaky iri'sa* ('coati-*onça*' or 'monkey-coati', since 'its hand is like the coati', mentioned earlier), *obaky poko* ('white *onça*'), and *obaky huj̃ huj̃* ('*onça coisa ruim*', literally 'bad-thing *onça*'). Note that the names of these beings refer to their qualities: to malice ('*coisa ruim*', 'bad thing', is a common name for the Devil in Brazil) and to what we might understand as a mixture of different beings including tree-dwelling animals, monkeys, and coatis. Conjunctions of oppositional animals like predators and prey (Oliver, 2015, p. 242) can signal ambivalences, while this association with animals that mainly live in trees is even stranger considering that they are opposed to *onças* in terms of high versus low: *onças* are the 'owners' (*ongy*) or 'chiefs' of all the prey animals that live on the ground, while the Harpy eagle (*boty'o*) is the owner of those that dwell in the canopy—birds, monkeys and coatis. These tree-dwelling *onças* therefore combine the attributes of high and low, of trees and the ground, as if collapsing the divisions that structure Karitiana ecology and hunting practices. And, as we know from Mary Douglas (1966), everything that transgresses categories must be distrusted.

But what are these practically cryptozoological creatures? As we have seen, these beings are *onças,* 'another type of *onça*', as the Karitiana would say. But it is precisely this monstrous aspect, the strange combination of different attributes of human and nonhuman beings, of high and low – the classic figure of the monster that appears to be universal or transcultural (Gilmore, 2003) – that allows us to understand the existence of these beings in the world inhabited by the Karitiana. This all points to the fact that what primordially defines Karitiana *onças* is not their morphology, anatomy, embryology, physiology, or their genes. Instead, they are a conjunction of affects; as Viveiros de Castro (1998) has formulated, bodies in Amerindian worlds are less material and more bundles of affects, conjunctions of dispositions that authorize the functioning of beings and their possibilities of action and reaction. In effect, these multiple *onças* are possible in the Karitiana universe because *onças* themselves are monsters, to a certain extent, or as the Karitiana say, *kida* (which they translate into Portuguese as *bichos*).

Bicho (*kida*) is a dynamic and interesting polysemic category. It is generally translated by the Karitiana as 'thing', as in *kida o*, 'fruit' (literally, 'round thing'). The term is used for specific creatures that are effectively monstrous and dwell in the forest or, on dark nights, in the villages. These include the *Mapinguari* (*owojo* or *kida harara,* 'laughing ogre') known in other Indigenous and non-Indigenous Amazonian locations (Vander Velden, 2016); the *Kida hu*ỹ *hu*ỹ (also known as *Te*ỹ *tema*), a monstrous winged beast, like a dog with bat wings, that descends to the ground with a canine whine to announce forthcoming deaths (note that this monster arose from an *onça's* tooth that *Boty*ỹ, the demiurge the Karitiana translate as God, painted and transformed into a *bicho*); the various forms of *psam'em*, pure white and aggressive cannibal spirits with bared teeth who wander confusedly in the neighbourhoods around cemeteries; the *Curupira* (*Dopa*), also called the 'mother of the forest' (see Figure 4.3); and *Ora*, a monstrous aquatic serpent, the owner of all beings that dwell in the water. But *kida*, the monstrosity, also refers to some animals with aggressive, predatory, or destructive dispositions towards humans: ticks and other blood-sucking animals, electric eels, scorpions, spiders, and *onças.*

The somewhat monstrous nature of *onças* can also be seen in alleged human metamorphosis. The Karitiana say that several techniques can transform a human person into a particularly fierce and bloodthirsty *onça*: drinking the blood of an *onça*; smearing *onça* fat on one's body; eating its raw meat; eating the leaves of a plant (*obaky irisap'*, literally '*onça's* ear'); eating the bark

FIGURE 4.3 *Dopa* (the *Curupira*), one of the Karitiana *kida* ('*bicho*', monster/beast).
SOURCE: DRAWING BY LUIS CARLOS KARITIANA REPRODUCED FROM ALTINI, 1999, P. 16

shavings of '*pau podre pintado*' (*ep'opokejema go obakydna*, literally, 'painted rotten stick'); or simply marking the body with the juice of the jenipapo fruit to imitate an *onça's* markings (*obaky ejep*', '*onça* painting'). The Karitiana resolutely refused to illustrate the jaguar's markings, even on a sheet of paper. The shaman Cizino once stated that painting the markings of an *onça* on paper would cause the animal to immediately jump out, putting our lives at risk, because the *onça* gets mad at anyone who imitates him, and consequently this is a '*coisa perigosa*' ('dangerous thing') to do. In the past, people used the jaguar pattern only in war and rituals, but they no longer do so today.

Moreover, eating the meat of jaguars makes a person 'heat up quickly:' in other words, to turn aggressive and violent (a 'hot' person's actions and emotions are uncontrolled). It is also said that an enraged person 'has no blood', like the souls of the dead. In the past, Karitiana warriors would make a tea from crushed and roasted *onça* teeth to 'get mad' for war. They reported that the danger of this practice was that the effect took some time to wear off, and could cause the warrior who drank the tea to turn against his fellow villagers: 'He would no longer want to know [human] persons or his family, he would be disgusted by people and would want to kill his relatives', says Valdomiro Karitiana—a classic result of a transformation into a jaguar, a creature that (by definition and principle in the South American lowlands) has no relatives, or does not respect them if it does. Note that the Karitiana never say that this human individual who becomes a jaguar views its relatives or fellow villagers as prey animals when he turns into a predator. Rather, they suggest that because this man-jaguar wants to kill his relatives, he does not recognize them, but he continues to see them as human persons and not nonhuman animals. It is clear that he views humans as prey just as an *onça* does, where he formerly saw human relatives; the point of view for this relationship has been altered.

'The *onça* is where *bichos* begin', Inácio Karitiana once stated; '*Obaky kida*' ('*onça* is a monster'), Epitácio asserted on another occasion, both during my fieldwork in the *Byyjyty osop aky* village. With this, they wanted to emphasize that these animals appear to exist in a zone between 'animal' and 'monster', between the everyday and the extraordinary, between the normal and the unheard-of. As suggested by Praet (2014), perhaps for the Indigenous Peoples of lowland South America, *onças* are what he calls *Monsters* before they are animals (*Beasts*). It is this ambiguity that permits the proliferation of the different types of *onça* that share the world with the Karitiana. They are attentive to the inherent breadth of behaviours, habits, forms, smells, sounds, and emotions of the *onças* of Amazonia: aware of them through many more aspects than those noted by biologists, since their very survival in the forest depends upon coexistence (peaceful or otherwise) with these wild animals. Thus, what to us may resemble cryptozoology among the Karitiana is converted into simple zoology. Perhaps the combination of *onças*, monkeys, and coatis in these beings points towards *onças* that frequent tree branches or share some sort of affect or behaviour with monkeys and procyonids. '*Onças-coisa ruim*' may refer to particularly large or singularly ferocious animals that attack people and frequently devour domestic animals; in the Pantanal of Mato Grosso, ranchers, in conflict with felines that prey on oxen, consider them pernicious animals (Süssekind, 2012).

What is notable is that *onça*-ness is not located in a particular zoological 'species' (*onças*, or jaguars), but rather dispersed across a myriad of beings that enrich the fauna of the world's largest tropical forest. A similar phenomenon has been seen in another region of Brazilian Amazonia among non-Indigenous river dwellers who also recognize a group of different (and non-scientifically recognized) *onças* (Macedo, Santos, Branquinho, & Bergallo, 2018, p. 130–135). Perhaps, from the Karitiana perspective, *onças* are becomings rather than beings, complexes of different but intimately related creatures which are very difficult or even impossible to separate into condensed entities: a characteristic of certain domains of microscopic life (Helmreich, 2009; Schrader, 2010) but one that appears to be shared by some so-called higher organisms that pose a cryptic challenge to the classificatory desire to locate each creature on an ordered list of discrete species (de Vos, 2017).

5 The Richness of the World

Could it be that the Karitiana recognize more types or species of animals than zoologists specialising in Amazonian fauna? This is a serious possibility, since they know the forest so well, and was confirmed a few years ago: in 2013, field researchers identified a new species of tapir ('a different type of tapir', as the Karitiana say) which the Indigenous People in that region had spoken of for quite some time (the black tapir, *Tapirus kabomani*; see Cozzuol et al., 2013). Or could the contrary be true; that the Karitiana confuse species and varieties of animals, identifying many types where only one or a few exist? Are the Karitiana experiencing the old ethnotaxonomical dilemma between superdifferentiation and subdifferentiation in the relationship between zoological species and native types (Berlin, 1973)?

Perhaps neither is true, and instead the Karitiana recognize beings of different ontological constitutions that interpenetrate and characterize an inherently dynamic taxonomy of animals and *bichos* (*kida*), in which beings are or become *bichos* only on certain occasions, as a result of certain affects and behaviours (such as being aggressive and attacking humans). Even though there is no idea of (and consequently, no word for) animal in the Karitiana language, as in other South American languages (Adelaar with Muysken, 2004, p. 234; Viveiros de Castro, 2006), this does not mean to say that this group does not recognize such a category (with different limits), even covertly, or that they perceive similarities in the beings that are opposed to humans, even though 'each nonhuman species is as different from all the others as it is from humans' (Viveiros de Castro, 2015, p. 226).

Onças are animals, but are also *bichos*: fundamentally ambiguous beings, and in fact located 'where the *bichos* begin'. They are nonhuman animals, charismatic predators, and also powerful owners of terrestrial animals. The different *onças* that live in the region where the Karitiana dwell thus demonstrate this oscillation between two categories of being, animal and *bicho,* and lead to a proliferation of types. The more that is known about a natural species, the more they are noticed, and the more the diversity of their forms emerges and becomes obvious. In a dynamic classificatory system like that of the Karitiana, the multiple differences are converted into types. The very concept of species as a recognized and stabilized artifact is placed in check ethnographically.

Hence, it seems that the three possibilities suggested by Forth (2017) for the proliferation of certain cats (several not recognized zoologically) among the Nage of the Island of Flores, Indonesia, could be operating in the Karitiana way of recognising the different *onças* (or, at least, more *onças* than those accepted by zoologists) with which they share their territory. Forth suggested the following options: first, the Nage have profound local knowledge of hidden species in the forest that biologists lack; second, fantastic beasts or imaginary beings are present in narratives concerning knowledge of local fauna (although the Nage, like the Karitiana, do not appear to attribute supernatural qualities to these felines); and third, super-differentiation is at play, the recognition of more species where only one is identified taxonomically, a differentiation based on wide variations in animal behaviour or morphology, which could account for the Karitiana way of recognising different *onças* (more than biologists and naturalists accept). However, none of these options alone can explain the Karitiana phenomenon: I suggest that an amalgamation of the three possibilities must be employed to understand the coati-*onças* and '*onças-coisa-ruim*'. At the same time, these beings are not recognized by science, since science does not yet seem to acknowledge the full range of possibilities of being *onça* in Amazonia, including those involving the supernatural – here not in the sense of the fantastic, but rather the monstrous.

In this way, following Hurn (2017, p. 211–212), these Karitiana *onças* could occupy a liminal position (a 'middle ground', in Hurn's words) between the creatures of experience (the real) and the creatures of the imagination, the unknown or the fantastic. They are transgressive creatures, to use Hurn's terminology again. This liminal space is often occupied by animals, that are recognized by science, when they exhibit anomalous behaviour that is either out of the ordinary or has not been described previously. As such, a world of experience opens up, focusing on ways of seeing, feeling, perceiving, and naming the incessant movement of things and beings. *Onças* do, and are capable of, many things: like other living beings, they are extremely rich in potentialities, affects, and almost unlimited gestures. Western zoology lists most of these potentialities as a set of habits, dispositions, or behaviours of one sole animal species, while the Karitiana seem to prefer multiple types, linking different actions or potentialities to different types of being. In the words of the Italian philosopher Emanuele Coccia (2016, p. 3), 'Every animal is a particular form of opening to the sensible, which is to say a certain capacity to appropriate and interact with images'. For this reason, I suggest that the nonhuman animals here are becomings: unlike in western zoology, beings are not 'detached from local ecology and most of the senses' and 'decontextualising nature' (Atran & Medin, 2010, p. 28–29). To the contrary, organising the discontinuities of the world depends on the experiences the Karitiana have with the beings they encounter 'embedded in the local environment', and which they observe doing ever-different things. As a result, they alter the way these beings are classified: a dog is not normally a *bicho* (*kida*), but if it attacks a passerby it could become one. This experience of the world, which preserves yet simultaneously and continuously transforms indigenous knowledge, is permitted specifically because the forests in Karitiana indigenous lands are relatively intact, with animals particularly abundant in the southern, eastern, and western areas farther from the villages.

Ultimately, the world is of a complexity and wealth so absurdly vast that no single human and no single society will ever be capable of understanding it in all its aspects (a point made by Graeber, 2015). Or as Gordon (2003, p. 6, my translation) aptly points out, 'The possibilities of classification and of description within this continuum that is nature are infinite'. The most that each society or culture can recognize of the world, and the things it contains, depends on the angle from which it views them: only some of the riches of the real world are illuminated, while many others are left in the dark (in zones that can become visible, depending on experience or experimentation with the real). This inherent complexity of life forms, and the natural fact that everything changes all the time, and never ceases to surprise us, permits its sheer variety to be perceived in multiple ways; each people, culture, or human collectivity delineates the real in its own way, favouring different manifestations or emergent forms. In the modern West, we have opted for stable categories through the basic concept of species: many differences were initially reduced to reproductive isolation (to species defined in terms of viable descendants), and in more recent times, to genetic maps. Meanwhile, the Karitiana seem to have opted for an inherent instability in the characteristics of living things, thus causing the proliferation of *onça*-ness in their forests as discussed here. There is not just one *onça*, but in a continuum between the house and the forest, between prey animals and beasts, between the natural and the supernatural, as it were. There are different *onças* of various forms, habits, aspects, trajectories, actions, and emotions. Therefore, animals are not just good to think about, to vilify, to prohibit, and to live with, but also good to multiply. These beings can also be called *chimeric* (*sensu* Severi, 2015), since they seem to combine attributes of different beings while at the same time presenting all of them in a single type: a coati-*onça* expresses and manifests both the *onça*/*obaky* and the coati/*iri'sa*. This 'presentification' indicates that the Karitiana are not using metaphors to connect distinct animals, but rather that these beings are part of their experience in the forest. Various people described having seen, encountered, or even slaughtered these different manifestations of *onça*-ness.

This is not an attempt to reduce one way of knowing the natural world to another: it is not a case of reaffirming the existence of one unique 'natural given' (the species *Panthera onca*), recognized by zoology, which the Karitiana multiply into several. At the same time, this does not mean to say that in fact many *onças* exist, and that modern science erroneously groups such different beings together in the same taxon; it is not that biologists are right and the Karitiana wrong, or vice versa. Instead, as a major challenge to anthropologists wishing to build bridges between Americanist ethology and anthropological studies of human-animal relations, this means recognising that both claims about *onças* and the natural world are correct on their own terms, that they make sense in their own historical and cosmological contexts and are ensconced within their own epistemological traditions. Finally, this situation entails acknowledging that natural beings, neither more-than-human nor less-than-human, but certainly beyond-human (in other words, different from humans), to a great extent lead their lives indifferent to humans and go about their business with no regard for indigenous or scientific taxonomies, even as the entire planet faces the tragic consequences of the Anthropocene era. To paraphrase Butler (2006), beings (like bodies) could be said to possess a history that is not only a social history. Perhaps my proposition here is a sort of material realism, with an assumption of the impermeability of certain objects to human cognition and also to knowledge (Pedersen, 2015).

Each researcher is a product of his or her own ontology; while it is pertinent to ask whether this immersion can be bypassed, including in the contemporary analytical reconfigurations in which natural and social/cultural are mixed in complex and apparently contradictory ways (Wolff, 2012), a perhaps more interesting question remains. If naturalism leaves a part of the world separated from human designs, do animists not do the same (supposing that the Karitiana can be described as animists)? Does this suggestion acknowledge that no one in this world knows everything or recognizes everything? Even if the animals' interiorities (souls) to Amazonian peoples are generally human, can it be supposed that the Amerindians know everything there is to know of the human phenomenon itself? Do they know everything of themselves? The experience of some anthropologists in the field, involving uncertain, tentative, tantalising, flagrantly invented or frankly mistaken responses to questions like these, would indicate that the answer is no.

It is precisely this independence of 'animal life' – which has been developing, specialising, and diversifying, for millions of years on this planet (before the presence of human beings and far from their activities), and which cannot ever be reduced to an anthropological or scientific object of study – that provides space for the emergence of diversity, for what we label biodiversity (used here without its quantitative connotations). Hence, as any fisherman knows, when human collectives throw their nets of thought and cognition over the natural world, they never know what they will catch – because of 'life's creative powers, of its inexhaustible power of transformation, of its plasticity' (Foucault, 1994, p. 126). The Argentine writer Jorge Luis Borges stated that hypotheses about the real must be interesting, while the real itself need not be (Borges, 2012, p. 159); perhaps the opposite is true for living things, where the real is so immensely rich and varied that even our most creative hypotheses will never come close to understanding it, nor prevent us from marvelling at all there is to continually surprise us.

References

Adelaar, W.F.H. (with the collaboration of P. Muysken) (2004). *The languages of the Andes.* Cambridge University Press.

Altini, E. (1999). *Universo cultural Karitiana.* CIMI-Regional Rondônia.

Atran, S., & Medin, D. (2008). *The native mind and the cultural construction of nature.* The MIT Press.

Berlin, B. (1973). Folk systematics in relation to biological classification and nomenclature. *Annual Review of Ecology and Systematics*, 4, 259–271.

Borges, J. L. (2012). La Muerte y la brújula. In *Ficciones.* Vintage Español.

Brown, C. (1999). *Lexical acculturation in Native American languages.* Oxford University Press.

Butler, J. (2006). *Gender trouble.* Routledge.

Coccia, E. (2016). *Sensible life: a micro-ontology of the image.* Fordham University Press.

Cozzuol, M., Clozato, C., Holanda, E., Rodrigues, F., Nienow, S., Thoisy, B., Redondo, R., & F. Santos (2013). A new species of tapir from the Amazon. *Journal of Mammalogy*, 94 (6), 1331–1345.

Descola, P. (1994). *In the society of nature: A native ecology in Amazonia.* Cambridge University Press.

Descola, P. (2005). *Par-delà nature et culture.* Éditions Gallimard.
Douglas, M. (1966). *Purity and danger.* Routledge.
Eisenberg, J., & K. Redford (1989). *Mammals of the neotropics – volume 1: The Northern Neotropics.* University of Chicago Press.
Eisenberg, J., & K. Redford (1992). *Mammals of the neotropics – volume 2: The Southern Cone.* University of Chicago Press.
Eisenberg, J., & K. Redford (1999). *Mammals of the neotropics – volume 3: The Central Neotropics.* University of Chicago Press.
Emmons, L., & F. Feer (1997). *Neotropical rainforest mammals.* University of Chicago Press.
Fausto, C. (2014). *Warfare and shamanism in Amazonia.* Cambridge University Press.
Foer, J. S. (2010). *Eating animals.* Little, Brown and Company.
Forth, G. (2017). Cryptids, classification and categories of cats: An ethnozoological study of unidentified felids from Eastern Indonesia. In S. Hurn (Ed.), *Anthropology and cryptozoology: Exploring encounters with mysterious creatures* (pp. 32–53). Routledge.
Foucault, M. (1994). *The order of things: An archaeology of human sciences.* Vintage Books.
Gambarini, A., Duarte, L., Haberfeld, M., & R. Cunha de Paula (2016). *Panthera onca: à sombra das florestas.* Avis Brasilis.
Gilmore, D. (2003). *Monsters: evil beings, mythical beasts and all manners of imaginary terrors.* University of Pennsylvania Press.
Gordon, F. (2003). Nossos Aipins são melhores do que os Outros. *Revista Habitus,* 1 (1), 15–32.
Graeber, D. (2015). Radical alterity is just another way of saying 'reality': A reply to Eduardo Viveiros de Castro. *HAU: Journal of Ethnographic Theory,* 5 (2), 1–41.
Helmreich, S. (2009). *Alien ocean: Anthropological voyages in microbial seas.* University of California Press.
Hurn, S. (2017). Land of beasts and dragons: Contemporary myth-making in rural Wales. In S. Hurn (Ed.), *Anthropology and cryptozoology: Exploring encounters with mysterious creatures* (pp. 203–217). Routledge.
Johnson, S. (2016). *Jaguar (Panthera onca) care manual.* AZA.
Kohn, E. (2013). *How forest think: Towards an anthropology beyond the human.* University of California Press.
Lévi-Strauss, C. (2004). *O cru e o cozido.* Cosac Naify.
Londoño Sulkin, C. (2005). Inhuman beings: Morality and perspectivism among the Muinane people (Colombian Amazon). *Ethnos,* 70(1), 7–30.
Macedo, J., Santos, R., Branquinho, F., & H. Bergallo (2018). Ameaça ou ameaçada? As controvérsias sobre as múltiplas onças na Amazônia. In V. Fernandez, J. Macedo, & F. Branquinho (Eds.). *Pedra, planta, bicho, gente...coisas: encontros da teoria ator-rede com as ciências ambientais* (pp. 127–150). Mauad/Faperj.

Maurer, B. (2000). A fish story: Rethinking globalisation on Virgin Gorda, British Virgin Islands. *American Ethnologist,* 27(30), 670–701.

Olivier, G. (2015). *Cacería, sacrificio y poder en Mesoamérica: tras las huellas de Mixcóatl, 'Serpiente de Nube'.* FCE/UNAM/IIH/CEMCA.

Papavero, N. (2015). *Sobre os nomes populares dados aos felinos do gênero Leopardus (Mammalia, Carnivora, Felidae).* NEHiLP/FFLCH/USP.

Papavero, N. (2017). Considerações sobre os felinos do Velho Mundo tratados como 'onças': notas históricas e etimológicas. *Arquivos do NEHiLP,* 14, 1–467.

Pavlinov, I. (2013). *The species problem ongoing issues.* Intech Books.

Pedersen, M. (2015). Islands of nature: Insular objects and frozen spirits in Northern Mongolia. In K. Hastrup (Ed.), *Anthropology and nature* (pp. 96–107). Routledge.

Praet, I. (2014). *Animism and the question of life.* Routledge.

Pratt, M. L. (1992). *Imperial eyes: Travel writing and transculturation.* Routledge.

Ritvo, H. (2018). Species. In: L. Gruen (Ed.), *Critical terms for animal studies* (pp. 383–394). University of Chicago Press.

Rosa, J. G. (1962). *Estas estórias.* José Olympio.

Santos-Granero, F. & G. Mentore (Eds.). (2006). Special issue in honour of Joanna Overing: In the world and about the world: Amerindian modes of knowledge. *Tipití,* 4 (1).

Saunders, N. (1998). *Icons of power: feline symbolism in the Americas.* Routledge.

Schmidl, U. (2008). *Viaje a España y las Indias.* Longseller.

Schrader, A. (2010). Responding to *Pfiesteria piscicida* (the Fish Killer): Phantomatic ontologies, indeterminacy, and responsibility in toxic microbiology. *Social Studies of Science,* 40 (2), 275–306.

Severi, C. (2015). *The chimera principle: an anthropology of memory and imagination.* HAU Books.

Süssekind, F. (2011). A onça pintada e o gado branco. *Anuário Antropológico,* 2011-II, 111–134.

Süssekind, F. (2014). *O rastro da onça. Relacões entre humanos e animais no Pantanal.* Editora 7 Letras.

Vander Velden, F. (2012). *Inquietas companhias: Sobre os animais de criação entre os Karitiana.* Alameda.

Vander Velden, F. (2016). Realidade, ciência e fantasia nas controvérsias sobre o Mapinguari no sudoeste amazônico. *Boletim do Museu Paraense Emílio Goeldi. Ciências Humanas,* 11 (1), 209–224.

Vander Velden, F. (2017). Como se faz um cachorro caçador entre os Karitiana (Rondônia). *Teoria & Cultura,* 11 (2), 25–35.

Villar, D. (2005). Índios, blancos y perros. *Anthropos,* 100 (2), 495–506.

Viveiros de Castro, E. (1998). Cosmological deixis and Amerindian perspectivism. *Journal of the Royal Anthropological Institute,* 4 (3), 469–488.

Viveiros de Castro, E. (2006). The crystal forest: Notes on the ontology of Amazonian spirits. *Inner Asia*, 9(2), 153–172.

Viveiros de Castro, E. (2015). *The native relative: Essays on Indigenous conceptual worlds.* HAU Books.

Vos, R. de (2017). Extinction in a distant land: The question of Elliot's bird of paradise. In D. B. Rose, T. van Dooren, & M. Chrulew (Eds.). *Extinction studies: Stories of time, death, and generations*, (pp. 89–115). Columbia University Press.

CHAPTER 5

Pilgrims and Other Sorts of Personifications: Nonhuman Animals as Ritual Participants in Isluga, Northern Chile

Penelope Z. Dransart

Abstract

Human and nonhuman animals go on pilgrimages as wayfarers, but usually only the human participants are accorded the status of pilgrim. Based on the author's ethnographic fieldwork among Aymara-speaking carers of llamas, alpacas and sheep in Isluga, northern Chile, this article demonstrates how herd members are co-participants along with human beings in ritual practices. Wayfaring as an activity is considered in relation to human personifications of other living beings, as well as of landforms in which the wayfaring takes place. Because the ritual inclusion of nonhuman animals takes place in a perceptual world that does not rest on an axiomatic split dividing ritual herders from non-ritual herded participants, the ethnographic evidence presented here indicates how dualistic principles in Isluga do not have the same contrastive force often imputed to Western ways of thinking and that commensurabilities between Andean and Western dualisms are partial, associated with different moral judgments.

Keywords

Aymara worldviews – human-nonhuman dualisms – personification – pilgrimage – relational ontologies – rituals of inclusion – wayfaring

1 Introduction

In the course of their daily herding tasks people in Isluga, in the Andean highlands of northern Chile, go wayfaring together with their llamas, alpacas and sheep. Undertaken on foot, wayfaring is a form of movement that also characterizes going on a ritual journey or pilgrimage. Human caretakers in Europe did not normally grant horses, donkeys and mules the status of pilgrim, even though these companions took them on their ritual journeying. Going

 | DOI:10.1163/9789004679450_006

on pilgrimage was allegedly a human prerogative. In European countries it formed part of a 'system of subjugation that differentially produces human and animal' (Steel, 2011, p. 237). In this exploration of who might be a pilgrim, I question the priority given to human beings. Can llamas, alpacas and sheep be considered, in some sense, as pilgrims, too?

Nonhuman beings in Isluga play significant roles in annual ceremonies designed to enhance the vitality of llamas, alpacas and sheep. Human and nonhuman pilgrims possess parallel but not entirely symmetrical identities. By co-opting llamas, alpacas and sheep into ritual life, people reveal different attitudes towards relationships between human and nonhuman beings than one might expect from a Western perspective. My aim here is not to erase boundaries between human and nonhuman, but to observe how practices in Isluga unsettle what Eduardo Viveiros de Castro (2014, p. 45) calls 'easy reductionisms' between being human and nonhuman.

Located next to the border with Bolivia, Isluga's extensive pasture grounds consist predominantly of scrub-like vegetation. Much of this steppe-like terrain is in excess of 4,000 m above sea level and, apart from some places where potatoes and quinoa can be cultivated, the conditions are not suitable for growing crops. This land nevertheless nourishes people and herds through cycles of scarcity and relative plenty. Through their daily and ritual practices, herders, llamas, alpacas and sheep know the terrain, its water sources, and the inconstant weather conditions.[1] Ritual activities arise from and are made meaningful by contexts occurring in the daily lives of human and nonhuman members of a pastoralist community. I therefore follow Gose (1994, p. 7), who states that a ritual 'depends on, and interacts with, other kinds of purposive activity' and Salomon (2018, p. 202), who sees religious experience as 'always welling up from ordinary life'.

Although motorized transport has replaced the use of llama caravans, Isluga people have for long journeyed from the highlands to the valleys and the coast in the west, as well as eastwards into Bolivia. Isluga forms part of the province of Tarapaca, territory that once belonged to Peru. During the War of the Pacific (1879–1883), Chilean forces occupied what is now the far north of Chile and the province has remained under Chilean jurisdiction since a proposed plebiscite in 1929.[2] Tarapaca's economy has relied on the extraction of minerals and

1 This article is based on fieldwork conducted in Isluga for periods of varying lengths of time in 1986, 1987–1988, 1989, 1995, 1996, 1997, 2003 and 2017. I am deeply grateful to the people of Isluga for making it possible to do the research reported here.

2 For a study of the social processes of Isluga's incorporation into the national history of Chile, see Ortega Perrier (1998).

nitrates, bringing in workers from far and wide – not only from the central belt of Chile, but also from Peru and Bolivia.

In a small desert community called La Tirana, 72 km south-east of Iquique, is a sanctuary dedicated to the Virgen del Carmen, based on a Marian cult introduced to the Andes by the religious order of St Augustine in the seventeenth century. Since the nineteenth century, pilgrimage to La Tirana has been expanding. It now takes place annually over a period of ten days, focusing on 16 July, attracting large numbers of pilgrims. The event is renowned for religious performances performed by the members of different confraternities who dance to the accompaniment of music in front of the church, in what is seen as a bewildering mix of traditional, religious and secularizing tendencies, bringing together people from different linguistic, ethnic and national backgrounds (Díaz Araya & Lanas Castillo, 2015).

Isluga's contribution to the pilgrimage has not been to join in the dancing. What Isluga people of various ages tell me is that herders have gone with their llamas and sheep to La Tirana to sell meat and to play music. People's attendance is for instrumental ends, which might be economic, and perhaps also to beseech divine assistance in the curing of illness. Mama Nati told me how she became a devotee of La Tirana after journeying there and the Virgen del Carmen restored her good health.

Isluga people's presence at La Tirana differs from that of other devotees. Their llama and sheep companions, however, will not return to Isluga after the pilgrimage. People are conscious that their herd animals sacrifice their bodies to sustain human life and their hides to cover the drums on which their caretakers beat the insistent rhythms to which other human participants dance.[3]

The language of the pilgrimage is significant, as highlighted in an invitation dated 1901. The priest of Camiña, a valley community with which Isluga people have long-established economic relationships, invited his parishioners to attend the fiesta at La Tirana '*en romerías*' ('in pilgrimage') (Díaz Araya & Lanas Castillo, 2015, p. 159). Bilingual Aymara- and Spanish-speaking people in Isluga apply the Spanish term *romero* or *romera* (a 'palmer' or 'pilgrim') to their llamas and alpacas in ritual song. In what follows, I explore some anthropological approaches to personification as a prelude to a consideration of how herding activities take place in terrain consisting of different levels of human and other-than-human interactions. I then turn to the ritual context in which people address nonhuman beings as pilgrims.

3 When speaking in Spanish, people use the verb *sacrificar* to refer to slaughtering a llama.

2 Knowing through Personifying

During an elaborate ceremony called the *wayñu* or *floreo* ('flowering'), which individual families conduct between January and March each year, people call their llamas, alpacas and sheep '*romero romero*' in ritual incantation. This personification goes beyond merely attributing personhood to nonhuman beings.

Sir Edward Burnett Tylor could have used llama pilgrims as an example to support his assertion that people in non-European parts of the world who appeal to personifications are 'animists' because of their 'deep-lying doctrine of Spiritual Beings' (1871, p. 384), but we (or whoever he imagined his readers to be) only do so 'in conscious metaphor' (p. 260). Michael Jackson observes that the contrast Tylor drew between non-European and European uses of personifications is congruent with the European Enlightenment's 'excoriation of anthropomorphism' and that vehement rejections of anthropomorphism served as a rhetorical strategy to conceal 'its pervasive presence in European discourse' (1998, p. 78). In such a light, Tylor's argument rests on the entirely problematic notion that people in a 'primitive' stage in the evolution of societies rely on conceptually simple and misguided perceptions. Tylor, however, did not deny the pragmatic force of metaphors; when users apply them, they define and redefine their experiences, as well as organizing and giving direction to ritual practices (Fernandez, 1986). Michael Taussig calls it a stand-off: 'Your metaphor. My literality' (1997, p. 4).

Isluga use of personifications reveals another contrast between European languages, on the one hand, and Aymara and Quechua in the Andes on the other. The former possess an overall term for 'animal', but in native Andean languages an equivalent category was not present and had to be borrowed from the Spanish (Adelaar, 2004, p. 234; Dransart, 2010, p. 89–90). Thus, an overall animal category, in which human beings may or may not place themselves (Ingold, 1994, p. 23), does not have the same emotive resonances.

Isluga herders tend not to speak of their llamas, alpacas and sheep using the borrowed Spanish term *animales*. They refer to them collectively instead as *uywa*, who are 'cared for' by human herders, a term sharing a linguistic root with *uywiri* ('those who grant existence'). *Uywiri* are topographical features such as hills and hollows in the land. People recognize them as the carers of vicuñas and guanacos (the wild forms of alpacas and llamas) and, in acknowledging the *uywiri* as the supreme herders in Isluga society, they are also the ultimate patrons from whom people 'borrow' llamas and alpacas (Martínez, 1976, p. 267–272). Because the *uywiri* are also *aviadores*, a term borrowed from the Spanish verb *aviar* 'to provide, to furnish', they are the 'providers' responsible for vitality in the multiplication of a person's herds (Martínez, 1976, p. 281).

These powerful landforms therefore present themselves to people as persons in a social environment comprising a hierarchy of herders. Kawaraya, a rugged snow-capped mountain, with his broad mouth, constantly watches over Isluga from his position in Bolivia. 'Why does he watch us and not Bolivia?' asked Mama Nati in my presence one day. In this respect, the hills in the region are like the 'watchers' Catherine Allen observed in Sonqo, Peru, 'against whom there is no concealment' (Allen, 2015, p. 25).

According to Tylor's logic, Andean landforms are personifications on the basis of his claim that people might believe something called spirit resides in them. Tim Ingold thought it better to consider how allegedly inert objects subsist in flux and transformation in a world inhabited by wayfarers, by persons who adopt different viewpoints through movement (2011, p. 29, 148).

Lest such a gaze prioritize human perception, it must be emphasized that Isluga people include *uywa* in the social group consisting of wayfarers; *uywa* join with herders to inhabit a world of personified landforms. Viveiros de Castro offers a way out the metaphor/literality impasse by taking into account 'the capacity [of a person] to occupy a point of view', which is a 'question of degree, context and position rather than a property distinct to specific species' (2014, p. 58). The relationship emerges from the interaction: 'One must know how to personify, because one must personify in order to know' (p. 62). This approach makes it possible to illuminate the character of mutual relationships between human and herd animals in Isluga and how other personified entities, who disclose themselves to herds and herders in the course of daily and ritual herding practices, can intermediate in their lives.

3 Rituals of Inclusion and Exclusion

As mentioned above, horses, mules, and donkeys went on pilgrimages in Europe, too, but their human keepers held the privilege of being the pilgrims. Esther Cohen (1994) recounted an eleventh-century tale of a French knight and his brother's mule, who together set out to go to Rome. The mule died near Conques where, in response to the knight's prayers, Saint Foy restored him or her to life. This miracle enhanced the reputation of the saint. Cohen (1994, p. 59) added that it also underscores the intimate closeness that was possible between human and nonhuman animals.

Cohen's study of legal processes and legal rituals in late medieval society of northern France, however, demonstrates how normative boundaries worked to establish a 'broad duality' in which 'the law of property considered animals chattels in every sense, [but] criminal law treated them as sentient, punishable beings' (1993, p. 101). Horses, mules and donkeys were ritually excluded as pilgrims. In contrast, Cohen argued that legal trials of animals facing criminal charges in a human court of justice served as rituals of inclusion, creating 'a universe of justice that transcended the human community' (1993, p. 100). Perhaps because of the bonds between them, animals were 'symbolic surrogates for people, a mirror held to humanity's face' (1993, p. 105). In practice she found that boundaries between human and other animals were not absolute because there were differences in the legal symbolism associated with ecclesiastical justice exercised by churchmen, in contrast to the attitudes of peasants and townsfolk who, in bringing charges against specific homicidal animals, did not regard all nonhuman creatures as belonging to a lumpen animality (Cohen, 1994).

A stark view of medieval violence against animals emerges in Karl Steel's (2011) study of what he called 'human limits'. A select few animals became cherished pets; for instance, he described how falcons belonging to the English monarch Edward I went on pilgrimage to shrines of the saints – in one case a wax image of a sick gyrfalcon was sent to Thomas Becket's shrine at Canterbury (Steel, 2011, p. 221). This treatment of a particularly intimate animal companion is based on 'the logic of the pet'. It did not necessarily involve the human keeper in protecting other members of the pet's species, nor did it 'demand that humans abandon their superiority to animals in general' (Steel, 2011, p. 222). Nevertheless, Steel thought it possible to find in the medieval sources expressions, however provisional, of symbiotic relationships between human and nonhuman animals that are not based on a 'closed human system' (2011, p. 240).

Published accounts of journeys made in the Andes for a religious purpose similarly do not recognize the ritual status of llamas, alpacas, donkeys, mules or sheep who accompany human pilgrims.[4] In pre-Hispanic times, people, llamas and alpacas went on pilgrimage to various holy places, one of them being the Island of the Sun in Lake Titicaca. Brian Bauer and Charles Stanish (2001) thought the dry windy season would have been the time of the year when most participants travelled to the island's Sacred Rock. They argued that the height of the activity probably coincided with the June (winter) solstice when the labour of agriculturalists was least needed. Their reconstruction of the event casts human beings in an exclusive role as pilgrims, with llamas and alpacas

4 It should be noted that whereas a donkey or mule may carry a human being on pilgrimage, the services provided by llamas, alpacas and sheep are different. Llamas could serve as burden-bearers but their ultimate role, as with alpacas and sheep, may be to provide a source of food.

featuring as sacrifices to divinities. In a seventeenth-century account, Ramos Gavilán listed the blood of one hundred camelids, accompanied by offerings of 'much wealth, like gold, silver, shell, and many feathers of various and very showy birds ... The animals to be offered were adorned with roses of various colors' (cited by Bauer & Stanish, 2001, p. 233).

4 Interacting Human and 'Other-than-Human' Worlds

A growing body of anthropological writings explore what or who a person may be in an interactive world of human and 'other-than-human beings'. These attempts provide useful grounding to counter the reductionism of treating human participants in a pilgrimage as pilgrims and nonhuman participants as non-pilgrims.

In the 1930s Irving Hallowell set out to the Lake Winnipeg region of northern Canada to do fieldwork among the Ashinaabe (Ojibwe). He encountered a perceptual world, which he regarded as a behavioural environment (a notion adapted from Gestalt psychology), populated not only by humans but also by 'other than human beings' (Hallowell, 2002 [1960], p. 21). The notion of what a 'person' was, he found, transcended the human category to encompass other phenomena in a moral universe in which it was possible for these 'other-than-human persons' to apportion their powers to their human counterparts (p. 46). In this view of the cosmos, such broadly conceived persons cause events to happen. Some rocks, for example, move or 'manifest other animate properties' (p. 24).

Nurit Bird-David (1999) responded to both Tylor and Hallowell in her exploration of how mutually constituted persons belonging to a larger-than-human category operate in Nayaka social environments in India. Beings called *devaru* become what she called 'superrelatives' because these beings engage with specific people, who encounter them in the course of their actions in the environment.[5] Like Hallowell (2002 [1960], p. 28), Bird-David (1999, p. S71) was uncomfortable with Tylor's use of the term 'supernatural' because it depended on antithetical concepts of nature and culture. Observing that Hallowell concentrated on human and more-than-human entities as categories, she emphasized how human and superrelatives' interactions in 'sharing

5 Bird-Davis rejected both 'spirit' as a translation of *devaru*, because it implies a 'spirit/body dualism of the modernist person-concept', and 'supernatural beings', which would mirror a Western idea of nature (1999, p. S71).

space, things, and actions' constituted persons (1999, p. S72; Naveh & Bird-David, 2014).

Eduardo Viveiros de Castro (1999) praised her observation that the personhood of *devaru* is revealed when they initiate their engagement with people and not vice versa. He questioned, however, whether Bird-David's use of the term 'superpersons', as a synonym for 'superrelations', projected the metaphor of a human person onto the 'super' quality of *devaru*. In response, Bird-David (1999, p. S87) explained that her interpretation of Nayaka ways of knowing the world through hunting and gathering activities was a 'specific cultural expression of a relational epistemology' as part of a more 'general human experience'. Her insistence that animisms exist in their diversity, there being no single form, therefore presents an important corollary to Ingold's point concerning subsistence in a world of fluxes and transformations (2011; see also Descola, 1996).

In conducting fieldwork among Anishinaabe in Lake Winnipeg country, Hallowell worked with people who had a foraging economy and had little exposure to Western modes of thought (1972, p. 58). In the absence of Western terminology with which to report his findings, he devised the phrase 'other than human', which has been influential in debates seeking to bridge natural and cultural domains. Alejandro Haber warns that, in discussions of relatedness, concepts such as the 'wild' and 'domesticated' qualities of nonhuman beings (he was referring to vicuñas and llamas) are easily reduced to hegemonic dualisms when employed as straightforward categories (2010). Closer examination indicates that European understandings of comparable ontological categories do not always collapse into hegemonic dualisms either.

Robert Brightman has characterized discussions on the relationship between nature and culture, and whether or not it is a cross-culturally valid distinction, as a struggle between monism and dualism (2007). His wry summary goes like this: at the end of the nineteenth century anthropologists claimed non-Westerners were in error because their animism prevented them from distinguishing nature from culture, only for twentieth-century French structuralists to decide that they were like Westerners after all in making such a distinction. Later in the twentieth century, anthropologists started to praise non-Westerners for their ecological awareness in not making the distinction. A consequence, Brightman argued, of this recent preference to see non-Western peoples as monists promotes a peculiarly Western nature-culture dualism to the status of a 'metalinguistic privilege', in turn prioritising 'the' Western form which, allegedly, everyone else 'universally and entirely' lacks or possesses (2007, p. 32).

Brightman uses this survey of changing universalist and relativist stances in the history of anthropological thinking to warn against using 'excessively categorical and dichotomous idioms [in which] whole societies are either dualistic OR monistic' (2007, p. 32). To avoid such a trap, in his fieldwork among Rock Cree, he explores partial commensurabilities, observing parallels and divergences between the accounts of his research participants and those of westerners (1993; 2007, p. 32).

In an effort to escape from an either-or impasse, Philippe Descola proposed a fourfold classificatory scheme based on the organisation of people's cognitive tendencies into 'dispositions of being' (2013). He takes two 'assets' into account, the first involving the sense of dispositions available to a person enabling him or her to act ('physicality') and the second a sense of inward self-reflection ('interiority'). His four 'routes' are:

Totemism:	mononatural in physicality and monocultural in interior disposition;
Naturalism:	mononatural in physicality and multicultural in interior disposition;
Animism:	multinatural in physicality and monocultural in interior disposition; and
Analogism:	multinatural in physicality and multicultural in interior disposition.

Descola finds totemism to be the predominant tendency in Australia, animism in lowland South America, North America, Siberia and some regions of South-East Asia, and analogism in China, other parts of Asia, West Africa, Mesoamerica and the Andes (2006, pp. 141–6). Instead of identifying naturalism with a geographical region, he considered it to characterize a chronological period, that of modernity. Some scholars accept that people's cognitive tendencies in Andean communities often express analogism (Salomon, 2018, p. 17, 52–57). Others do not; they draw instead on literature on animism (Sillar, 2009; Allen, 2015; Stensrud, 2016) or they seek to open analytical paths by displacing Westernized grammars of nature (de la Cadena, 2015).

Descola does not claim that his 'dispositions' constitute mutually exclusive ontologies, although he suggests that animism and totemism are strongly opposed to each other (2006, p. 146). Another way of looking at the problem is to regard his strategy as being in itself an overarching analogous act. In his schematic patterning, the four categories of person understand their environment by 'inferring analogies and contrasts between the appearance, behaviour, and properties' in encounters with other entities (human and nonhuman) (Descola, 2013, p. 112). It is therefore difficult to disentangle his 'analogism' from the other three dispositions. Catherine Allen maintains that in the Andes analogical modes of social relationships rely on underlying animist tendencies (2016, p. 331).

Bororo cognitive tendencies, in Eastern Brazil, provide an example of co-present modes of identification based on animic and totemic forms of relatedness between humans and others. In Crocker's (1985) analysis, animals are either *aroe* or *bope*. Members of the felidae family, for example, belong to the *aroe* category and are totemic. Among *aroe* there are also spirits possessing no 'material analogue in this world' (Crocker, 1983, p. 165). In contrast, *bope* are associated with life processes, exchanging vital energies with humans.[6] This example of co-present ontologies is intriguing because, as explained in the next section, Gabriel Martínez wondered if the different settlements in Isluga territory had a totemic relationship with specific *uywiri* (1976, p. 276, note 12).

5 Interspecies Interpellations

Martínez's account makes the *uywiri*-hills act like *aroe* in Crocker's description of Bororo totemism.[7] Adult Bororo men could assign fifty or sixty named *aroe* to each clan, but they only considered some five or six to be 'great' (Crocker, 1983). In Isluga, all hills have the potential to reveal themselves as *mallku*, but not all *mallku* are *uywiri*-hills and only certain prominences gain recognition as the nurturer of a specific settlement (Martínez, 1976, p. 272). Bororo held certain aspects of *aroe* personification to be 'publicly secret', a term that Crocker used in relation to the concealment of phenomena that are 'sacred' in Durkheimian usage – 'set aside, hidden, mystically potent, opposed to mundane secularity' (Crocker, 1983, p. 175, note 2). Martínez noted that hills in Isluga have a 'common' name in parallel to an 'almost secret' one (1976, p. 273).

The most powerful of these hills are at the top of the trophic order because they are capable of 'eating' human and nonhuman alike, a commonly invoked metaphor when people talk about the hills surrounding them (Dransart 2022). Mechanisms for what Signe Howell calls 'feeding relationships' express significant social interactions (1989, p. 429). In Isluga, the special potency of the metaphor derives from the pastoralist context and important ritual activities

6 Crocker, writing at a time before animism began to regain a popular place in anthropological conversations, placed more emphasis on what he regarded as Bororo totemic epistemology rather than see the metonymic chains of association between *bope* characteristics and landforms or creatures of natural worlds as a form of animism (1985, p. 127).

7 That Martínez was writing in the 1970s and Crocker in the 1980s makes this parallel more striking.

are intended to promote the health of pasture grounds where llamas, alpacas and sheep feed.

In this world of personified landforms, people and *uywa* together cultivate their terrain in a domain of interpersonal relationships sustained by the *uywiri* and the Wirjin Tayka ('Virgin Mother'), the powerful being known as Pachamama in other parts of the Andes (Dransart, 2002a, p. 55). People seek to '*dar fuerza*' ('to give strength') to the land so that these beings can impart vitality to human and herd animals (Martínez, 1976, p. 297).

These landforms seem to act like Bird-Davis's 'superrelatives'; they do more than just watch because they intimate their existence in people's lives, just as particular stones made themselves known as 'grandfathers' in Hallowell's account of the Ashinaabe. Certain standing stones in Isluga are called 'grandfather', too (Figure 5.1). Such a stone is a survivor from a previous world age who has revealed his presence by making noises, as though playing the drum (Dransart, 2002a, p. 61). Some rock formations are female in gender – Q'asiri shouts out to the unwary herder or passing llama.[8] Isluga people often give more salience to the gendering of these beings instead of explicitly identifying their human or nonhuman characteristics.[9] The stones need not be large. Nazario, a resident from the Peruvian community of Pacchanta, in the cordillera of Ausangate, told Marisol de la Cadena of an *inqaychu,* in this case a small, alpaca-shaped piece of the mighty mountain Ausangate, 'that Ausangate had made my father find' (2015, p. 107). The important point here is that these powerful beings, whatever their dimensions, participate in people's lives.

Personified stones are witnesses to a previous world age inhabited by undomesticated people (Isluga people refer to them using the Spanish term *gentiles* or 'pagans'), in which guanaco and vicuña were the bearers of culture in a land lit only by the light of the moon (Dransart, 2002a, p. 61; 2002c, pp. 61–62). Elsewhere, in Sonqo, Peru, Allen was told that in those apocalyptic times herbivores ate their human herders (1988, p. 98).

This widespread myth in the Andes suggests a prior world order in which nonhuman animals led a civilized way of life. Brightman pointed out that in Néhinaw (Cree) traditions humans obtained cultural possessions from other animals and he juxtaposed a Bororo and a Asinīskāwiyiniwak (Rocky Cree)

8 For a discussion of the 'character' of personified landforms as animated beings see Dransart (1996; 2002a, p. 59–64).

9 There is a marked emphasis on female-male dualisms in Isluga and my discussion returns to dualisms below. See also Martínez (1976) and Mamani Mamani (2010). The counterparts of the lords *mallku* referred to in this article are the ladies *t'alla* – Q'asiri being a fierce example of the latter.

FIGURE 5.1 A grandfather stone standing at the edge of raised ground in the midst of moist pastures between Enquelga and Caraguano, settlements in Isluga, northern Chile
PHOTO: PENELOPE DRANSART, 2017

myth. In the former, Jaguar came to hate humans more than other living beings because his adopted son stole his stepfather's fire, forcing him to renounce grilled food, as well as his bow and arrow, compelling him to rely on his fangs to hunt. All that remained was the fire's reflection in his eyes. In the Asinīskāwiyiniwak myth, Wolverine lost his fire along with its plume of smoke. Lest all the other animals' smoke plumes draw people to hunt them in the future, he persuaded them, too, to renounce the use of fire.

Brightman (2007, p. 37) commented that among Asinīskāwiyiniwak and Bororo, these accounts presuppose the dualistic contemporary environment to be the product of nonhuman animals' renunciation of 'the arts of civilization'. Therefore, he argued, the division between the cultured human domain and the natural nonhuman domain has less contrastive force than in a Western form of dualism. A consequence is that such a Asinīskāwiyiniwak perspective does not take human rational achievements necessarily to be superior to those of nonhumans, although human customs have a 'malleability and transmissibility' in contrast to the 'more 'necessary' – and thus more 'natural' qualities ascribed to the behavioral repertoires of different animal kinds' (Brightman, 2007, p. 39). This idea of dualism not necessarily having contrastive force lends itself to different judgments concerning moral hierarchies between human and nonhuman beings. It sets the scene for considering the ritual context in which llamas and alpacas are treated as pilgrims.

6 Mutually Constituted Ritual Persons in Isluga

Ceremonies intended to enhance the fertility of the herds take place during the rainy season in Isluga, when llamas, alpacas and sheep join human herders as ritual participants. One of the places where I observed these ritual events was in corrals owned by families on rising ground in an area of moist pasture with a standing stone grandfather nearby, one with a reputation for being particularly noisy. Isluga people call these ceremonies *wayñu*, which literally is a form of dance, or *floreo*, because it is a 'flowering' of the herd animals. Each family in the community organizes its own event sometime after 1 January and before the moveable date of Carnival (it begins on the first Saturday occurring before the start of Lent). If a family cares for sheep, their ceremony will be held after Carnival.

The ceremony opens with an all-night vigil during which people sing songs and make ritual libations, followed by daytime events in a corral. One of the herd is selected for sacrifice and libations are made over his or her back. To the accompaniment of music that participants acknowledge as sad sounding, the other llamas and alpacas are ritually dressed with brightly coloured ear tassels and neck-pieces. Men cut notches in the ears of young llamas, alpacas, and sheep who are approaching sexual maturity and women stitch the tassels through the ears of all the animal participants. All humans present tie fleece, which has previously been dyed red, orange or yellow, to the withers of female animals and along the length of the spine of male animals (Figure 5.2). At the end of day, the ritually reinvigorated llamas and alpacas or sheep go out to the hillsides. The human participants return home to share a communal meal and to dance to the sound of joyful music.

Llamas and alpacas cry and people incessantly sing plaintive incantations to the accompaniment of music during the ceremony. Some ethnographers of the Andes have admitted to me (privately) that they find the ritual songs of the communities of northern Chile, including Isluga, to be repetitive and lacking in poetic value. Yet Manuel Mamani Mamani, a native of Guallatire in the Chilean highlands north of Isluga, stresses that each ritual event in the annual cycle 'possesses an exclusive text, melody and rhythm' (2006, p. 45). Isluga people say their family's songs have a different *tono* ('tone') to those of their neighbours, and each family asks permission of their own *uywiri*-hill to proceed with their *wayñu*. During the event, incantations and mandolin music

FIGURE 5.2 In the centre, a male *jañachu* alpaca has been ritually dressed during a *wayñu* ceremony with rose-red *chimpu* fleece tied on his head and along the spine, ear tassels and a neck-piece of pendant cords.
PHOTO: PENELOPE DRANSART, 1987

provide the animating force to ensure that the rites will be effective. Alcoholic drinks are sprinkled on the backs of the animals and on the earth, the body of Wirjin Tayka. Through their alcohol moistened throats, men and women sing over and over, directing their voices to *uywiris* and to *uywa*, including the *romero romero* incantation, hailing the latter as pilgrims.

In February 1997, this song resonated through the community. After a family sung it to their llamas in the corral, it spread to another family whose members no longer celebrate the *wayñu* because they have converted to an evangelical protestant cult. Converts are not allowed to drink alcohol and, as one man said in sorrow, 'they do not love the *uywiris*'. Members of this evangelical household were caring for a frail grandmother who, four days after the celebration in the corral, I heard singing *wayñu* tones. The sadness of the songs made her cry, perhaps because they brought back memories of llamas and alpacas who did not live their allotted time-span, lost to illness, accidents or vengeful acts committed by *uywiri*. At times she was confused because she said she was singing for carnival, to which non-evangelical members of the community were now preparing to go. Her granddaughter responded by singing repeatedly the *romero romero* song to her grandmother, keeping it alive in her mind.

The sung incantations convert llamas, alpacas and sheep into pilgrims whose respective journeys through life towards death are sustained by the reinvigorating forces of Wirjin Tayka. Because *uywa* enter the corral with depleted fertility and they emerge from it in a revitalized condition, people perform the *wayñu* in all its elaborateness in an attempt to protect against sterility. Good fortune will ensure that fertility brings blessings enjoyed by the herd animals and, by extension, their fortunate human herders.

In this ritual context, people use the Spanish term *bendición* when libating liquids or powdery substances to the land (Dransart, 2002b). Hubert and Mauss (1899) recognized the importance of libations because they provide a means to disperse energized droplets of liquid into a ground of depleted vitality. Making libations is not only a human task in Isluga, because llamas, alpacas, and sheep also have a participatory role in this ritual work. People use the Aymara term *phawaña* ('to sow') when dispersing dry substances.[10] One of the contexts in which they apply it refers to the brightly hued tassels and dyed fleece known as *chimpu* that fall from *uywa* during the lapse of time from one *wayñu* ceremony to the next (Figure 5.3). Hence llamas, alpacas, and sheep take an active part

FIGURE 5.3 A brightly hued tassel, considered to be ritually 'sown' (*phawaña*) in the pasture ground by the llama or alpaca from whose ear it dropped
PHOTO: PENELOPE DRANSART, 2017

10 I am grateful to Juan de Dios Yapita for clarifying the meanings of *phawaña*.

in making ritual offerings because they sow the 'flowers' with which they have been ritually dressed during the *wayñu* during the course of the annual cycle of the seasons.

In Isluga, women weave belts with a white zigzag line along each longitudinal edge, immediately inside the outer selvedges. They call these line *cerros* ('hills'), the domain of the wild partridges whose eyes occur in the repeated diamonds that constitute the central axis of the belt. Arnold, Diego and Espejo have discussed zigzag designs in contemporary Andean textiles, associating them with pathways called *q'iwi q'iwi* in Aymara, in reference to a 'movement, usually of a bird or animal, but in a zigzagging, abstract pattern' (2014, p. 295–7). The contours of hills require the creatures of this world to take back and forth pathways, which are presented graphically on the belts worn by Isluga men, women, and children. Zigzag lines also remind weavers of kinked intestines, which they trim when butchering a llama or alpaca. When I showed a photograph to an Isluga weaver of sharply zigzagging rock formations I had seen in the Quebrada de Humahuaca, Argentina, she was fascinated, commenting, 'those rocks are like intestines'. In the course of their herding activities, people and *uywa* experience how rock formations make them take sharp turns in their wayfaring into a world populated by birds and other beings. After the *wayñu*, *uywa* and people go forth into that world ritually empowered as pilgrims.

7 A Final Discussion

This consideration of llamas, alpacas and sheep as pilgrims exemplifies that human beings are not always the sole participants in ritual practices. The ontological grounding underlying the co-participation of people, *uywa* and *uywiri* in daily and ceremonial life demonstrates the persuasive power of personifications amongst Isluga pastoralists.

Pilgrimage in various parts of the world has conventionally been seen as a human affair. Taking into account different motivations and the types of journeys pilgrims make for sacred purposes, Alan Morinis (1992, p. 10–14) devised a six-fold typology. Five of these types (devotional, instrumental, normative, obligatory and initiatory) are characteristic of pilgrimage made by human beings in the Americas – for example Juan van Kessel (1992) laid great emphasis on the instrumental character motivating pilgrims to visit shrines in northern Chile for health reasons. Pilgrim *uywa* in Isluga also participate on this instrumental basis because the *wayñu* initiates them into sexual maturity or revitalizes the older members of the herd. Morinis and Crumrine (1991) argued that pilgrimages in Latin America always have a sacred goal and that the sixth type, wandering pilgrimage (without a geographic destination), does not occur. *Uywa*, however, are different from their human counterparts. Although some llamas and sheep accompany people to La Tirana, most of them spend daily life in wandering pilgrimages, where their wayfaring takes a zigzagging course through the pasture grounds of Isluga.

The *wayñu* is an exceptional event when human and nonhuman participants experience heightened sensations. It enables them to constitute themselves mutually as ritual persons in relation to *uywiri*-hills. The Inka custom of adorning llamas and alpacas with 'roses', seemingly a precursor to the rose-coloured *chimpu* used in Isluga, hints that contemporary relational practices have emerged from deep-rooted traditions. Relational modes of knowing, however, must have differed in Inka times under a bureaucratic authority that exerted control over human and nonhuman beings, somewhat in the manner of a nation state today having the authority as well as the capacity to issue every domestic herd animal within its borders with a passport.

If there are commensurabilities between 'Western' and 'Andean' systems of thought, they are partial. This article has demonstrated that axiomatic splits in nature-culture or human-animal dualisms do not cleave cleanly as they are supposed to do, even in Western style traditions inherited from the Enlightenment. Brightman considered the continuation into the Age of Reason of the European practice of committing animals to law courts to be a phenomenon undermining the official Christian dualistic separation of human from nonhuman ways of being. A plausible explanation, he argued, was that people subjected pigs and other animals to courts of law because many of them really did believe in a 'subdominant doctrine of ensouled animals' (Brightman, 2007, p. 40). He declared a corollary: if Europeans never were consistently thorough in their dualistic worldviews that the human system was hermetically sealed to nonhuman beings, we should not be surprised if Asinīskāwiyiniwak or other peoples simultaneously hold different propositions regarding their relatedness with other beings of the world they inhabit.

In the examples from the Americas presented here, myths and rituals tend to downplay strongly contrastive dualisms. Culture, once associated with nonhumans, became the property of humans among Asinīskāwiyiniwak in whose keeping it acquired a greater sense of malleability and transmissibility than had been the case when kept by animals. The Isluga *wayñu* ceremony is a rite of inclusion because it incorporates people and *uywa* as ritual practitioners; human and nonhuman beings alike are responsible for cultivating their environment. People sing to their llamas, alpacas and sheep using a pilgrim metaphor that has, using James Fernandez's term, enactive force rather than just symbolic value (1986, p. 7). Nonetheless, humans take on a heavier load of ritual work than they expect *uywa* to perform.

Brightman (2007, p. 41) was confident that the authority of dominant monisms and dualisms would not replace local ways of knowing, but would interact with them, producing new forms based on changing environmental experiences. Undoubtedly, Isluga people, llamas, alpacas and sheep will continue to respond to new ideas interactively, based on their experiences of a perceptual world that does not rest on an axiomatic split dividing ritual herders from non-ritual herded participants.

References

Adelaar, W.F.H., (with collaboration of P.C. Muysken) (2004). *The languages of the Andes*. Cambridge University Press.

Allen, C.J. (1988). *The hold life has: Coca and cultural identity in an Andean community*. Smithsonian Institution Press.

Allen, C.J. (2015). The whole world is watching: New perspectives on Andean animism. In T.L. Bray (Ed.), *The archaeology of wak'as: Explorations of the sacred in the pre-Columbian Andes* (pp. 23–46). University Press of Colorado.

Allen, C.J. (2016). Stones who love me: Dimensionality, enclosure and petrification in Andean culture. *Archives de sciences sociales des religions 61e année*, 174, 327–346.

Arnold, D. Y., Diego, M. de & E. Espejo (2014). On the relation between Andean textile iconography and woven techniques. In D. Y. Arnold with P. Dransart (Eds.), *Textiles technical practice and power in the Andes* (pp. 283–302). Archetype.

Arnold, D. & J. de D. Yapita (1998). *Río de vellón, río de canto: cantar a los animals, una poética andina de la creación*. Hisbol/ILCA.

Bauer, B.S. & C. Stanish (2001). *Ritual and pilgrimage in the ancient Andes: The Islands of the Sun and Moon*. University of Texas Press.

Bird-David, N. (1999). 'Animism' revisited: Personhood, environment, and relational epistemology. *Current Anthropology*, 40 (Supplement), S67–91.

Brightman, R. (1993). *Grateful prey: Rock Cree human-animal relationships*. University of California Press.

Brightman, R. (2007). Nature and culture in the bush: A nature/culture metalanguage and Rock Cree parallels. In F. Laugrand and J.G. Oosten (Eds.), *La nature des esprits dans les cosmologies autochtones. Nature and culture of spirits in Aboriginal cosmologies* (pp. 31–44). Les Presses de l'Université Laval.

de la Cadena, M. (2015). Earth beings: Ecologies of practice across Andean worlds. Duke University Press.

Cohen, E. (1993). *The crossroads of justice: Law and culture in late medieval France*. Brill.

Cohen, E. (1994). Animals in medieval perceptions: The image of the ubiquitous other. In A. Manning and J. Serpell (Eds.), *Animals and human society: Changing perspectives* (pp. 59–80). Routledge.

Crocker, J.C. (1983). Being an essence: Totemic representation among the Eastern Bororo. In N.R. Crumrine and M. Halpin (Eds.), *The power of symbols: Masks and masquerades in the Americas* (pp. 157–176). University of British Columbia Press.

Crocker, J.C. (1985). *Vital souls: Bororo cosmology, natural symbolism, and shamanism.* The University of Arizona Press.

Descola, P. (1996). Constructing natures: Symbolic ecology and social practice. In P. Descola and G. Pálsson (Eds.), *Nature and society: Anthropological approaches* (pp. 82–102). Routledge.

Descola, P. (2006). Beyond nature and culture. *Proceedings of the British Academy*, 139, 137–155.

Descola, P. (2013). *Beyond nature and culture*, translated by Janet Lloyd. University of Chicago Press.

Díaz Araya, A. & P. Lanas Castillo. (2015). Danza y devoción en el desierto: Obreros e indígenas en la fiesta de la Virgen del Carmen de La Tirana, norte de Chile (siglo xx). *Latin American Music Review*, 36(2), 145–169.

Dransart, P.Z. (1996). Las flores de los rebaños en Isluga: La vida cultural de los ganaderos y camélidos en el norte de Chile, *Nuevo Texto Crítico*, Year 9, 18, 29–39.

Dransart, P.Z. (2002a). *Earth, water, fleece and fabric: An ethnography and archaeology of Andean camelid herding.* Routledge.

Dransart, P.Z. (2002b). Concepts of spiritual nourishment in the Andes and Europe: Rosaries in cross-cultural contexts. *Journal of the Royal Anthropological Institute*, 8(1), 1–21.

Dransart, P.Z. (2002c). Coloured knowledges: Colour perception and the dissemination of knowledge in Isluga, Northern Chile. In H. Stobart and R. Howard (Eds.), *Knowledge and learning in the Andes: Ethnographic perspectives*, (pp. 56–78). Liverpool University Press.

Dransart, P.Z. (2010). Animals and their possessions: Properties of herd animals in the Andes and Europe. In M. Bolton and C. Degnen (Eds.), *Animals and science: From colonial encounters to the biotech industry* (pp. 84–104). Cambridge Scholars Publishing.

Dransart, P.Z. (2022). Comer y ser comido: fisiologías de la práctica en Isluga, norte de Chile. In L. Bugallo, P. Dransart and F. Pazzarelli (Eds.), *Animales humanos, humanos animales. Relaciones y transformaciones en mundos indígenas sudamericanos* (pp. 355–386). Editorial Antropofagia.

Fernandez, J. W. (1986). *Persuasions and performances: The play of tropes in culture.* University of Indiana Press.

Gose, P. (1994). *Deathly waters and hungry mountains: Agrarian ritual and class formation in an Andean town.* University of Toronto Press.

Haber, A.F. (2010). Animismo, relacionalidad, vida: Perspectivas post-occidentales. In Darío Hermo and Laura Miotti (eds), *Biografías de paisajes y seres. Visiones desde la arqueología sudamericana*, (pp. 75–98). Facultad de Humanidades de Catamarca and Encuentro Grupo Editor.

Hallowell, I. (1972). On being an anthropologist. In S.T. Kimball and J.B. Watson (Eds.), *Crossing cultural boundaries: The anthropological experience* (pp. 51–62). Chandler Publishing Company.

Hallowell, I. (2002 [1960]). Ojibwa ontology, behaviour, and world view. In G. Harvey (Ed.), *Readings in indigenous religions* (pp. 17–48). Continuum.

Howell, S. (1989). Of persons and things: Exchange and valuables among the Lio of Eastern Indonesia. *Man*, N.S. 24(3), 419–428.

Hubert, H. and Mauss, M. (1899). Mélanges d'histoire des religions: essai sur la nature et la fonction du sacrifice *Année sociologique*, II, 29–138.

Ingold, T. (1994). Humanity and animality. In T. Ingold (Ed.), *Companion encyclopedia of anthropology: Humanity, culture and social life* (pp. 14–32). Routledge.

Ingold, T. (2011). *Being alive: Essays on movement, knowledge and description*. Routledge.

Jackson, M. 1998 *Minima ethnographica: Intersubjectivity and the anthropological project*. The University of Chicago Press.

Kessel, J. van (1992). Faith healing in the north Chilean Andes of pilgrimage. *Exchange*, 21(1), 62–83.

Mamani Mamani, M. (2006). Tierra, producción y música en el área agrícola de Parinacota: El caso del rito de Pachallampi. In P. Dransart (Ed.), *Kay pacha: Cultivating earth and water in the Andes* (pp. 39–46). Archaeopress, BAR International Series 1478.

Mamani Mamani, M. (2010). Kirkir warmi: Identidad y rol de la mujer aymara en el desarrollo musical del norte chileno, *Revista Musical Chilena*, Año LXIV, No 213, 90–102.

Martínez, G. (1976). El sistema de los uywiris en Isluga. *Anales de la Universidad del Norte Homenaje al Dr. Gustavo le Paige, S.J.* 10, 255–327.

Morinis, A. (Ed.), (1992). *Sacred journeys: The anthropology of pilgrimage*. Greenwood Press.

Morinis, A. & N.R. Crumrine (1991) La peregrinación: The Latin American pilgrimage. In N.R. Crumrine & A. Morinis (Eds.), *Pilgrimage in Latin America*, (pp. 1–17). Greenwood Press.

Naveh, D. & N. Bird-David (2014). How persons become things: economic and epistemological changes among Nayaka hunter-gatherers. *Journal of the Royal Anthropological Institute* (N.S.) 20, 74–92.

Ortega Perrier, M. (1998). *By reason or by force: Islugueño identity and Chilean nationalism*. [Unpublished PhD thesis]. University of Cambridge.

Salomon, F. (2018). *At the mountains' altar: Anthropology of religión in an Andean community*. Routledge.

Sillar, B. (2009). The social agency of things? Animism and materiality in the Andes. *Cambridge Archaeological Journal* 19(3), 367–377.

Steel, K. (2011). *How to make a human: Animals and violence in the Middle Ages*. The Ohio State University Press.

Stensrud, A. (2016). Climate change, water practices and relational worlds in the Andes. *Ethnos* 81(1), 75–98.

Taussig, M. (1997). *The magic of the state*. Routledge.

Tylor, E.B. (1871). *Primitive culture: Researches into the development of mythology, philosophy, religion, art, and custom*, Vol 1. John Murray.

Valeri, V. (2013). Marcel Mauss and the new anthropology. *Hau: Journal of Ethnographic Theory* 3(1), 262–286.

Viveiros de Castro, E. (1999). Comments. *Current Anthropology* 40 (Supplement), S79–80.

Viveiros de Castro, E. (2014). *Cannibal metaphysics*, P. Skafish (Ed. and translator). University of Minnesota Press.

CHAPTER 6

The Fragility of Relations of Domestication: Humans, Llamas, and Unseasonal Snow in the Bolivian Andes

Maggie Bolton

Abstract

In June 2002 heavy snow fell unexpectedly in the dry season in the south of Bolivia. This chapter looks to the occasion of this snowfall, and its disastrous consequences for llamas and their herders alike, to consider the materiality of human-animal relations in the Andes. It takes up Ingold's call to consider life as unfolding in a 'weather world' rather than a landscape, and Anderson's focus on 'architectures of domestication' that considers domestication to be a material and spatial arrangement. In examining a case in which the weather rendered places normally convivial to humans and herd animals hostile, it concludes that domestication itself is a fragile arrangement that can be ruptured. A focus on the materiality of domestication takes into account the presence of human bodies and raises questions about the viability of Andean herding in an era of increased human migration, when extreme weather events appear to occur more frequently.

Keywords

domestication – herd animals – materiality – weather – Bolivia

1 Introduction

> In the time of our grandfathers, there were large numbers of llamas, but there was no freezing weather. The weather was normal, and the rains were sufficient. There was water in the irrigation ditches. But now, at the time in which we're living, the rains don't fall at the right time. The cold spells are also arriving very early and for these reasons our animals are gradually dying off.
>
> ANONYMOUS LIPEÑO HERDER (*Lowe 2004*)

 | DOI:10.1163/9789004679450_007

In this chapter,[1] I draw on fieldwork conducted in the normally arid highlands of Sud Lípez province in southwest Bolivia, and focus on an unseasonal weather event, a snowfall in 2002, which proved disastrous for many Lipeño llama herds. My aim is to bring into studies of Andean pastoralism recent scholarship on human-animal relations, including on the material arrangements of domestication, and anthropological writing on weather – on the common immersion of humans, nonhuman animals and their environments in what Ingold terms a 'weather-world' (Ingold, 2007, 2010, 2011).

Recent work on human-animal relations, particularly from northern pastoral societies, has focused on the 'architectures' or infrastructures of domestication, that is, on the physical structures associated with confining animals, places in the landscape convivial to both animals and humans, and the tools used in animal husbandry (Anderson, 2014, Anderson, Loovers, Schroer & Wishart, 2017). While there are differences between the trajectory of domestication and pastoralism in the sub-Arctic and the Andes – llamas are different from reindeer, not least in their resilience to snow – this line of thinking opens a new perspective on Andean herding. The inclusion of weather, the medium in which architectures, humans and animals are all immersed, situates the chapter within a recent trend to take anthropology beyond the human, which is particularly pertinent, since my aim is to show how human-animal relations can be transformed by adverse meteorological events. I focus on the weather at a time when, according to herders, extreme weather events are becoming more frequent and when Lipeños, for economic motives, are managing their herds in non-labour-intensive ways. I argue that this focus highlights the entanglements and dependencies of humans and nonhuman animals in places of domestication and, importantly, the fragility of these arrangements.

2 Humans and Llamas in Sud Lípez: a Historical View

The present-day province of Sud Lípez came into being in the nineteenth century, but the name Lípez is older, being the name that sixteenth-century Spaniards associated with the high, arid region. Early accounts of Lípez by colonial Spaniards focus on four topics: the province's 'barbaric' people, mineral wealth, livestock, and the weather, although Spaniards were far more preoccupied with minerals than they were with either weather or llamas. Examples include Juan de Betanzos' *Narrative of the Incas*, which mentions the province's livestock (Juan de Betanzos, 1987 [1557], p. 164), and the letter of the *Factor de Potosí*,

1 A longer, Spanish-language, version of this chapter is published in L. Bugallo, P. Dransart & F. Pazzarelli (Eds.) (2022).

Juan Lozano Machuca, (Lozano Machuca, 1992 [1581], p. 30), which describes its mines. Luis Capoche in his *Relación general del asiento y Villa Imperial de Potosí* comments on the weather and the harshness of the land:

> It is cold and dry, and fierce winds always blow. It rains little and is uninhabitable save to the barbarous nation that populates it, people with neither order nor governance. It has extremely high mountains, perpetually snow-capped, and salt flats where no crops or grass grow. In the foothills of its mountains lie the settlements of its Indians who survive on roots, quinoa and a few potatoes, with no other means of sustenance.
>
> CAPOCHE, 1959 [1585], (p. 127)

While the salt flats and agricultural activities are present in what now forms Nor Lípez province, the cold, dry, windy weather-world with its high mountains evokes present-day Sud Lípez – although the snow caps on the mountains are no longer present all year.

Nielsen presents zoo-archaeological evidence that Lipeño people at the time of the European invasion were specialized pastoralists, independent of larger agro-pastoral vertical polities (2009, p. 22, 30). Indeed, this independence may explain why the Spaniards considered its people so barbaric. While Nielsen (2009, p. 22) considers their motivation for camelid domestication to have been meat, fibre and the animals' load-carrying capacity, Dransart (2002, p. 11–12), writing on the basis of archaeological and ethnographic evidence from Isluga in Chile, puts the emphasis firmly on fibre. In so doing, she draws attention to the pre-eminence of weaving among Andean technologies and to differences between animal domestication in the Andes and in the pastoral societies of the Old World.

While a detailed history or archaeology of camelid herding in Lípez is beyond the scope of this paper, in acknowledging recent scholarship positing domestication as a mutual process (see Cassidy & Mullin, 2007), we can presume that the relation between humans and llamas benefits llamas in extending to them a degree of protection from predators, while humans gain access to fibre, meat and the animals' capacity for carrying loads. Indeed, during colonial times, the employment of llamas to carry minerals from the Lipeño mines may even have encouraged the proliferation of herds in Lípez. Apart from transporting minerals, herders would have used llamas for journeys of exchange like those that present-day herders recall and that some ethno-archaeologists have reported: to the salt flats in the north and the valleys of the warmer parts of Lípez, Chichas and Tarija – their most documented journeys being to exchange salt for maize (see Nielsen, 1997, p. 98). By travelling together humans and llamas experienced and came to know the weather-world of Lípez, and beyond,

FIGURE 6.1 An elderly woman herder watches her llamas before the annual animal-marking ceremony
PHOTO: M. BOLTON, 2004

in each other's company. Platt, in his study of nineteenth-century Lípez, notes that herders prioritized these journeys, and the social relations and access to produce that they enabled, over earning cash from transporting minerals (Platt, 1987, p. 522).

Journeys of exchange have all but ceased with the proliferation of motorized transport. This change has inevitably altered the knowledge humans and animals have of their world and has led to routes and paths falling into disuse. Nevertheless, Lipeño herders still acknowledge the centrality of llamas to their lives (Bolton, 2000, p. 173–4). Few have ceased to keep animals, even if, with new economic demands and opportunities, they now migrate to Bolivia's cities to study or work. Responsibility for herds often falls now to older women: the grandmother of a family will remain at an isolated homestead while the younger generations are absent. I have heard such grandmothers declare that they have no wish to leave and want to die with their animals. Indeed, in the cold months of the Lípez winter, each year a few *abuelas* succumb to hypothermia.

3 The Weather and the Andean *Ayllu*

In many accounts of Andean societies, we read about the prevalence of the *ayllu* as a form of human corporate landholding collectivity (Salomon, 1992,

p. 21), and of relations between humans, nonhuman animals and others unfolding in an animate Andean landscape comprising powerful beings who watch over people. Prominent among these powerful beings are the mountains themselves (e.g., Allen, 1988; Martínez, 1983; Harris, 2000; Ødegaard, 2011).

Scholarship has moved on since anthropologists wrote of Andean mountains as landscape deities, and anthropologists have used a variety of theoretical perspectives, ranging from animism to actor-network theory (Latour, 1993), to try to understand the relations that prevail between rural Andeans and their surroundings (e.g., Sillar, 2009). In a post-human interpretation from southern Peru, de la Cadena (2014, 2015) writes of rural *ayllu* members (*runa*) as 'not-only' human. Her understanding is that they form part of *ayllus* that comprise people, certain animals, powerful nonhuman earth-beings (*tirakuna* Quechua- 'places') like the mountains, and the land itself. All these categories of being, as de la Cadena (2014, p. 255) states, make the place that they also are, the *ayllu*. In her formulation, earth-beings or *tirakuna* are with runa in *ayllus* and cannot be disentangled from each other (de la Cadena 2014, p. 255).

Wild animals are cared for by the earth-beings who also care for humans. Those domestic animals considered to be *uywa* (Martínez, 1989, p. 26) – that are raised by humans in a relation analogous to, and nested within, that which people have with earth-beings – are entangled with people and places. Llamas come into this category, and if relations unfolding within the Andean landscape recall Latour's actor-network theory, these relations with animals in-*ayllu* also resonate with Despret's (2004, p. 131) visualisation of animal domestication as an articulation of 'with-ness'. Llamas are domestic animals, different from the wild camelids of the Andes, and llamas, humans and places are inextricable from each other, forming an arrangement of llamas-with-humans-with-places. This entanglement is illustrated by a Lipeño myth, which posits llamas as animals original to the land – Lipeños contend that llamas originally emerged from a *ciénega* (marsh) in the foothills of Cerro Lípez. In fact, they consider llamas as more connected to place than they are themselves, since, while claiming to be *originarios*, they relate family histories of migration from elsewhere to work the mines. We might thus say that Lipeños, llamas, wild animals and earth-beings together form the *ayllus* of Sud Lípez.[2]

2 The *ayllus* of Sud Lípez were recuperated around the turn of the millennium after falling into disuse following the Bolivian National Revolution of 1952. Although actors involved in the recuperation had political motivations, recuperating the *ayllus* was important to Lipeños generally in reflecting the relation they understood between themselves and the place of which they form a part.

Anthropologists write of the Andean world as an animate landscape (Sillar, 2009; Harris, 2000), but Lipeños also speak of the weather as animate. Lightning and thunder are powerful entities, winds are associated with illnesses, and rains that are supposed to arrive in November are associated with the dead, since their arrival coincides with that of the souls for Todos Santos. In attending to their formulations, and following Ingold (2011), we should possibly speak of a sentient weather-world, rather than landscape, for this conveys something about life unfolding in a medium, rather than things upon the surface of the earth (Ingold 2011, p. 126). Ingold argues that we, other living beings and non-living things, are all related through common immersion in a medium that surrounds us, shapes us, and in which we experience the world (Ingold, 2011, p. 129). The interfaces between earth, water and sky are constantly being formed and transformed. 'Weather-world' attempts to convey that we are in a world that is in constant movement, flux and becoming (Ingold, 2011).

4 Architectures of Domestication, the Weather-World and Lipeño Herding

Recent writing on domestication has critiqued earlier studies that focused on the agency of human beings and human domination. Lestel urges us to engage with hybrid human-animal communities in studying how humans and animals live together (Lestel, Brunois & Gaunet, 2006) and Despret sees the practices of domestication as 'practices that create and transform through the miracle of attunement' (Despret, 2004, p. 125). She reminds us that domesticates are 'animals-with-humans', just as their associated humans are 'humans-with-animals'. In similar vein, Haraway suggests we 'become-with' our companion animals (Haraway, 2008, p. 16–17). Turning to herd animals, Istomin & Dwyer (2010), characterize the domestication of Siberian reindeer as a dynamic mutual adaptation of behaviour. They critique approaches that see domestication as resting on human agency and use ethnographic observations to demonstrate both the mutuality of behavioural modifications that underlie domestication and the recursive nature of such modifications (Istomin & Dwyer, 2010, p. 620).

In an unpublished short note, Olivia Harris suggested that 16th-century Andeans may have conceived of their *ayllus* as being 'rather like llama herds, and their *mallku* or *jilakata* [leader] as a herder' (Harris, 2009). If humans and llamas have (be)come together through dynamic mutual accommodations and if humans indeed modelled their organisation on that of their herds, then, recursively, their herds' organisation was always already an outcome of the animals' association with humans and human interventions. From fieldwork in contemporary Isluga, Chile, Dransart (2010) shows how Andean herders acknowledge the agency of their domestic camelids. Their herding practices do not so much impose human domination as work with the animals as acting subjects. They expect their llamas to have agency or autonomy and to develop qualities like leadership appropriate to their kind (Dransart, 2010, p. 91).

Lestel, Brunois & Gaunet, (2006) acknowledge that material artifacts underlie hybrid communities, but do not discuss them in detail. Earlier writing described tools of domestication as 'tools of coercion' (e.g., Ingold, 2000, p. 307). However, Anderson, Loovers, Schroer & Wishart, (2017) take a more nuanced approach, pointing out that although some tools can be used to inflict pain or force, they are infrequently used thus. In exploring human-animal relations in northern societies, the authors posit domestication as a spatial relationship involving an 'architecture' (see also Anderson, 2014), comprising not only built structures, such as corrals for confining animals, but also the tools used by humans, and convivial places, attractive both to humans and the animals they herd (Anderson, Loovers, Schroer & Wishart, 2017, p. 401). They argue that the frugal architecture of domestication of northern reindeer herders is self-limiting – enabling some human control over animals, but also encouraging mutual autonomy of reindeer and people. The instruments used by reindeer herders effectively define a setting where humans and animals can work together or collaborate and help define spaces that are good for both (Anderson, Loovers, Schroer & Wishart, 2017, p. 400).

Anderson characterizes domestication as a collaboration between humans, animals and landscapes, and the tools used by humans as enabling a setting in which humans and animals can cooperate (Anderson, Loovers, Schroer & Wishart, 2017, p. 406). Nevertheless, thinking through experiences in Lípez, two other factors contribute to settings in which humans and animals can come together. One is simply the material presence of humans, pertinent in Lípez now that so many people are absent from their herds. The other is the weather. We should, perhaps, speak of the architecture of domestication as including features not of the landscape, but of the weather-world, since the spaces that are good to live in – convivial for humans and animals – comprise human-made structures and tools, features of land and water, and the medium in which all are immersed – the changing atmosphere that forms their weather.[3]

The architecture of domestication is as sparse among Lípez llama herders as among northern reindeer herders. Herding tools are limited to sacks, llama-fibre ropes and slings, and human-built structures – homesteads which comprise simple adobe dwellings for humans, and dry-stone walled corrals for animals.

3 Sara Schroer, one of Anderson's co-authors, discusses the weather in the sections of the jointly authored article dealing with human-bird relations in falconry (Anderson, Loovers, Schroer & Wishart, 2017).

Some herders confine their female llamas with young in corrals at night – particularly during the warmer rainy months from December to February, when young are born – and sometimes set a slow-burning dung fire at the entrance to deter predators. Others bring the animals into the corral only occasionally, for the annual llama-marking ceremony held before Carnival, or when a veterinary procedure, such as vaccination, is to be performed. Ropes are used to close the corral entrance and were formerly used to attach sacks of produce to the animals' backs on journeys of exchange. Slings are used when driving animals, to lob a stone in front or to the side of the lead animals, to encourage them to turn in the desired direction. Vocal commands are also used (see also Dransart, 2010, p. 96), but their effectiveness depends upon the wind and its direction.

With this scant architecture of domestication – self-limiting, as in the Siberian case – it is unsurprising that physical contact between herders and animals is minimal. In former years, castrated male llamas had more contact with humans through their use as beasts of burden. Now, caring for llamas largely involves watching them, and watching out for them. Humans watch over llamas, as they consider themselves to be watched by 'earth-beings', the high mountains (c.f., Allen, 1988, p. 40–41). Herders make sure they know where their animals are grazing, sometimes shepherd them in an appropriate direction, make sure none gets lost and try to keep them safe from predators. Largely, however, the animals exercise autonomy in that they decide for themselves where they will roam and the company they keep – herders acknowledge that certain animals that are friends can always be found together.

Most watching by herders takes the form of brief daily checks, often involving a long walk from the village or homestead. Women and children (when not in school) sometimes spend longer with their animals, the woman sitting spinning or knitting while her herds graze. The practice of keeping male and female animals in separate herds that ethnographers recount from other areas (e.g., Stobart, 2006) is not enforced in Lípez, and may have been abandoned as people sought non-labour-intensive ways of herd management. Today, female llamas graze with their young and the occasional full male (*janachu*) on low-lying pastureland, while groups of young and castrated male animals often wander further afield, in groups that mirror the troops of young male vicuñas (wild camelids) that frequently graze downstream from San Pablo.

In the Lípez weather-world, interfaces between air, water and ground are sharp in some places, but can become blurred and indistinct. Earth and water mix in the moist *bofedales* where llamas graze and the saturated *ciénegas*, where humans harvest the alga *llullucha*. There, the surface of the land becomes unclear as solid and liquid intermingle. From the north of Bolivia, the Kallawaya understanding of the body and its relation to the environment, described by Joseph W. Bastien, anticipates Ingold's ideas. Bastien reports how the air shared by people, plants and animals is understood as a unitive principle between them (Bastien, 1985, p. 599). In Lípez, air penetrates the earth in the mines, and what is land and air becomes unclear when the feet of people and llamas, as well as the occasional passing vehicle, kick up dust that becomes airborne. Sand is also whipped up by strong winds that cause it to accumulate in loose drifts (of sand and air) on the old road to Pululus – now impassable to vehicles – and even in the streets of San Pablo, the province's capital.

While air may be animating, and a unifying principle, as Bastien (1985) describes, and the primary medium in which humans and animals are immersed, as Ingold (2011) notes, movement of air, or wind, is also dangerous. Bold, writing from the north of Bolivia notes that in Kaata wind is a living entity that can carry effects from afar – the damaging effects of climate change in the case she researched (Bold, 2019, p. 107). Lipeño people talk of illnesses being winds. An oft-recounted myth tells how a plague that destroyed the colonial mining centre of San Antonio del Nuevo Mundo was personified by a woman, María Picha Picha (Bolton, 2006a). It took me a while to realize that in recounting how she danced, entering in turn into each of the settlement's houses, my Lipeño narrators were describing the movement of wind.

On winter nights, the temperature falls well below zero and even summer nights, when the air is moist, are not exactly warm. Llamas are more resilient than humans to being in the open – although they suffer in the rainy season, usually between December and February, as their coats are not very water resistant. Outside these months the weather is normally dry, and, in the winter (June, July and August), the diurnal temperature range is high. During the day, under clear skies and the intense mountain sun, the temperature can reach over 20°C, but it drops rapidly below zero after sunset. This is the time of year when, formerly, herders would journey to the valleys. Today the animals are used for family subsistence – llama meat is sent to family members in the cities – and, with the growing popularity of the meat among urban consumers, both meat and *ch'arki* are sold in Tupiza. Even then the weather world intervenes, making the road to Tupiza impassable when sand accumulates on the mountain pass, the Paso del Diablo. Women weave less than in former years, but llama fibre is still used for making blankets and ropes, and some is sold.

This, then, is the Lípez weather-world where humans and nonhuman animals live together. As Ingold notes, common immersion in the 'generative fluxes of the medium' (Ingold, 2007, p. S19) is what unites humans and animals and enables relations between them. The normal predictability of the weather helps with the loose herding arrangements that Lipeños now prefer. Herders are content if their animals graze in the places they crave, which are also comfortable for humans, and where they require little attention. These are the places of domestication. Things change, however, when extreme weather events occur, when the places of the weather-world become hostile and human-animal sociality is imperilled.

5 The Snowfall of June 2002

Occasionally, usually in years when oceanographers note the fluctuations in the Pacific Ocean temperature termed ENSO (El Niño Southern Oscillation), unseasonal weather occurs. Most often, Lipeños complain of drought, and they can cite the years of severe water shortages. They associate such droughts with El Niño, which is widely reported in the Bolivian media. In such years, if they slaughter a llama, they invariably complain that the animal is thin. Less frequently, unseasonal snow falls in the dry season. Usually little falls, and it is no more than an oddity, evaporating rapidly under the intense solar radiation at high altitude. Very occasionally the snowfall lasts longer, to drastic effect. This is what happened in 2002 and has recurred on a couple of occasions since – although with less devastating consequences.

While light snow falls in the warmer rainy season, it does not often settle, although it usually leaves snow-caps on the high mountains that persist into the dry season. Lipeño people watch the weather constantly and now post about it on social media, along with complaints about the government. When they see clouds, they speculate on whether, and how much, it will either snow or rain, and they anticipate unseasonal snow much more frequently than it actually falls. In my first winter in Lípez, my landlady seemed constantly to be predicting snowfalls that never arrived. However, in 2002 heavy snow did fall. It fell for four days in June, causing devastation and damage that was felt for months, and even years, afterwards. I missed the snow, being in the UK, but was present for the distribution of aid to its victims some weeks later and I witnessed its consequences in the lives of herders over the following two years. I then realized why snow was so constantly in the minds of Lipeños.

The snow of June 2002 was unlike rainy season snow, in that it persisted on the ground and, with the ever-present wind, in places formed deep drifts. Lipeño people retreated into their dwellings, their town houses in San Pablo or their *campo wasis* (out-of-town homesteads), relying on stored foodstuffs and firewood to cook and keep warm. One friend, a *concejal* (municipal councillor) who had a thermometer, told me that the temperature fell below −20°C in his rented room in San Pablo. For days people were separated from their animals and in some cases their human relatives. The Bolivian media was captivated by the story of a child found with frostbite in her fingers and toes. She was rescued by a Peruvian helicopter crew near the Quetenas villages after she tried to walk home with her two cousins from their *campo wasi*. The two cousins perished.[4] Four humans died of hypothermia or frostbite during the snowfall and, according to a FAO report, 26 500 llamas perished across the four provinces of Sud Lípez, Nor Lípez, Antonio Quijarro and Tomás Frías. This number represented some 14% of the total domestic camelid holdings in the wider region, with Sud Lípez being the worst affected area (FAO Bolivia, 2009).

Wild as well as domestic animals died as a result of the snowfall, including vicuñas, the wild camelids present in large numbers in Sud Lípez. However, it was the loss of domestic llamas that affected Lipeños greatly and which they lamented. The animals formed an important part of their livelihoods, but beyond this they were also the animals to whom they were supposed to offer protection but had been unable.

As herders, Lipeños are accustomed to animal deaths, but not all deaths are equal. Like pastoralists elsewhere, they slaughter no more animals than is necessary, and there is an accepted manner in which slaughter is performed. One does not slaughter one's own animals but calls upon the services of a male neighbour or relative to hold the animal down, slit its throat and perform most of the butchering. The man in return receives his *achura* (Bolton, 2002, p. 394), in the form of some ribs and a piece of the liver to take home. Animals are carefully selected for slaughter – by age and gender – in accordance with requirements for the well-being of the herd and the needs of their owners.[5] Although herders always comment on whether the animal was fat or thin at the time of slaughter, the occasion is happy, as humans feast afterwards on *morcilla* (black pudding) and roast meat, dogs drink the leftover blood, and there will be meat for soup in the weeks that follow.

Occasionally, an animal meets an untimely end. This was the case with Trenzani, a llama named for markings on her neck resembling a woman's plaits, who became unable to stand and, consequently, had to be slaughtered. The cause of her malady – a broken rib – became clear when she was butchered. This was a bad death, for Trenzani was a young female with reproductive potential. Furthermore, she had been the only llama belonging to the daughter

4 See https://www.bolivia.com/noticias/autonoticias/DetalleNoticia7422.asp (Retrieved May 22, 2020).

5 I have retained the word 'owner' here to differentiate the person to whom an animal is assigned from those who provide care on a daily basis, who may or may not be the same. Lipeños describe themselves as owners of animals, either in Spanish or through using possessive suffixes in Quechua. However, this is shorthand for the more complex, nested relation of care that I describe in this article.

of my host from her previous marriage. The girl, already considered a poor orphan following her father's death, became materially poor with the loss of her last animal. Like Trenzani's untimely end, the deaths of animals in the snowfall were bad deaths. The snowfall killed llamas with no regard for the wellbeing of the herd or the welfare of the animals' owners. As Convery, Baily, Mort & Baxter (2004) noted, for farmers in the English Lake District following the 2001 foot-and-mouth epidemic, death occurred in the wrong place, at the wrong time and on the wrong scale. While Despret (2004) notes how trust and attunement come together in arrangements of domestication, the Lípez snow impeded attunement and betrayed trust. The effects of the snow on llama herds, llama mortality and the health of surviving animals became constant topics of conversation in the months that followed.

Not being llamas (or llamas-with-humans), we can only try to empathize with them and imagine what the snowfall was like for them. Following Ingold and focusing on the medium in which they were immersed, we can think of how the normally dry air would have become moist and mixed with water as the snow began to fall and land on their coats, and then how the surface of the ground must have become ill defined, as it first became damp and then covered with a thick layer of water-with-air as snow settled. Llamas lack lanolin in their coats to repel moisture, and so the cold wet snow settling on them, rather than providing insulation, would have made them feel the cold and become susceptible to hypothermia.

In time, not only would familiar features of the weather-world have altered but also, as the snow deepened, the *ichhu* grass on which they depend for food would have disappeared beneath the white blanket. Llamas have soft feet and tread delicately on the earth – Lipeños and NGO technicians alike will tell you that they do not drag their feet like sheep – and they eat by nibbling the tips of *ichhu*, rather than by pulling plants out of the soil, as sheep do. With the snowfall – at best – they were unable to find food and were at risk of starvation as the snow persisted on the ground. In the worst cases, animals floundered, becoming disoriented and hypothermic. Some, such as those belonging to my friend don Mauricio, were understood to have made a (fatal) mistake in choosing where to walk. Their owner told me how they became trapped where the snow settled in drifts and were buried. Snow makes the ground uncertain and when animals or humans cannot feel the ground upon which they depend, they must rely on relations with others (personal communication, Tim Ingold). In this instance, their relation with humans did not help them.

Lipeños lamented the loss of their animals. The snowfall happened when NGO technicians had been trying to convince them that there were commercial markets for llama products. In response, herders had started to increase their herd numbers. Since llama reproduction is slow, and Lipeños believe that increase should come from within their herds, by multiplication (Urton, 1997, p. 70–73), rather than by importing animals from outside (Bolton, 2006b, p. 541), time and effort had already been committed. It was thus hard for herders, in some cases, to lose half their animals and see their efforts go to waste. One woman commented that it was made harder by the meat from the animals dying in the snowfall being unfit to eat – she drew attention to its discoloration – so that the carcasses had to be buried, sometimes at the instigation of NGOs.

In the following weeks the snow gradually melted, the media arrived and departed again, and aid from national and international NGOs arrived. This aid took the form of tools, water purification kits, animal fodder (alfalfa and oat hay), and animal medication. In the months that followed, pasture was scarce –*ichhu* had died and rotted under the snowy blanket. Surviving animals went hungry and the rotten pasture gave them diarrhoea. Any llama slaughtered for meat that year was inevitably reported as being 'thin'.

6 The Rupture and Re-establishment of Relations of Domestication

> When there was the heavy snowfall last year, it almost finished off our llamas and sheep. They were out in the corral and the snow fell and fell. The llamas could not get up. They were buried and for want of help they died. Perhaps it is because we have forgotten how to make the offerings to the mountains, that our grandparents used to make, that this suffering has come to Lípez, since these freezes have been happening for a few years now.
>
> ANONYMOUS LIPEÑO HERDER, (Lowe, 2004)

In discussing domestication, its emplacement and associated architectures, Anderson (2014) emphasizes the idea of 'home creation', '… the identification of special places in the landscape that both animals and people crave' and how, 'the physical setting drives people and animals into a co-existence creating a mutual interest in creating a common life together' (Anderson, 2014, p. 19). While not denying that domestication involves a degree of coercion and control over animals, Anderson emphasizes the mutuality of a relationship that involves a coming together of people and animals in convivial places. Stepanoff, (2017) commenting on Anderson, Loovers, Schroer & Wishart, (2017), appreciates this spatial approach, but considers the account somewhat idealized in that it does not take into account failures and ruptures. He notes that, for reindeer herders, each year many animals abandon domestic herds and become wild (Stépanoff, 2017, p. 417). Domestication is not forever: it can fail or be reversed.

The snowfalls in Lípez resulted in the normally convivial places in the landscape becoming hostile and dangerous. As the quotation above demonstrates, even the interior of corrals, structures built by humans to keep animals safe, became dangerous when the snow fell and drifted until the animals could no longer get to their feet. This represented a rupture or failure in the relationship of domestication: a situation in which humans were unable to keep animals safe.

Ethnographers of domestication in the Andes contrast *uywa*, animals cared for by humans, with *sallqa*, those cared for by other beings that Lipeños sometimes equate with the mountains and sometimes with ancestors (Pazzarelli, 2022; Martinez, 1983; Dransart, 2010). Pazzarelli (2022) notes how these are not exclusive categories, but are nested, such that an animal raised by humans is ultimately cared for by the mountains, since these care for the humans themselves. This leaves open the possibility that animals can become undomesticated if they revert to the care of the mountains. Dransart (2010, p. 90) contends that the domestication of herd animals in Isluga, Chile is an ongoing activity, the result of daily caring activities, and that each generation of herd animals must be brought into a state of domestication by herders. Her observation recalls Platt's (2001, p. 652) findings regarding humans in Ayllu Macha in northern Potosí – that new people are conceived in an original pagan state and only after birth can a new-born be converted into a small Christian being. Each new generation of people must be brought into a state of domestication through post-partum rites of separation from the mother. Platt (2001, p. 652) writes of the 'voracity' of the pagan foetus and how this voracity must be 'domesticated' in the new-born, by curtailing its access to colostrum from the mother's breast, which is considered 'too rich'. Pazzarelli (2022, p. 306–10) notes a similar approach to animals, in Jujuy, Argentina. There, the access to alfalfa permitted to young goats is managed almost obsessively to curb their voracity. Without this control, his respondents claim, they would 'become wild' and the mountain would become their owner.

Lipeños contend that if one does not take proper care of one's animals, then the mountains, or ancestors, can take them back. Sometimes 'taking the animal back' involves the animal's death as when some affliction occurs, like *sarna* (mange). *Sarna* is held to be a punishment from the ancestors. Before antiparasitic baths and injectable medicines were available, the only way to combat the condition was to kill the animal and place its bones in the river (Bolton, 2010, p. 181). Death and reversion to the wild both involve the loss of the animal to the herder. The failure to keep llamas safe from the snowfall led to the loss of many through death. In some parts of Potosí department, I have heard that herders were so upset at losing their animals in the snow that they did not want to return to herding (personal communication, Juan Villanueva). I do not think Lipeños even considered that course of action, but they had to work to re-establish shattered bonds of domestication with their herds.

In the weeks following the blizzard, aid organisations brought supplies to Lípez for humans and animals. While Lipeños were pleased to receive donations, their reactions to some goods, particularly animal fodder, perplexed NGO technicians. The fodder was intended as a short-term measure to feed animals while pastures recuperated, but herders were reluctant to use it. One NGO reported that herders did not find barley straw acceptable, but would accept alfalfa and oat hay. While they accepted the fodder, most families hoarded the bales rather than feeding it to their llamas although, following the arrival of the trucks, many llamas wandered into San Pablo and helped themselves to the straw that littered the ground. The reluctance to use the fodder was not a case of saving it for later. Herders explained to me that having eaten such rich food, the llamas would not want to eat local *ichhu* again. In other words, recalling Pazzarelli (2022) and Platt (2001), they would become too voracious. Bringing llamas back into a state of domestication involved curbing their desire to eat and re-acclimatising them to the Lípez weather-world.

Relations with the environment were also renewed through ritual. A public response was the revival of a ritual ascent of Cerro Bonete by the men of Lípez communities, to make sacrifices to the mountain. Bonete is regarded as the most important and powerful mountain in the region, yet the ascent had not

FIGURE 6.2 Following the snowfall of 2002, a donation of animal fodder is received in San Pablo de Lípez
PHOTO: M. BOLTON, 2002

been carried out for some years. It had been neglected during the period following the National Revolution when *ayllu* organisation had been abandoned temporarily. The motivations for the revival of the ascent were complex. In addition to concern for re-establishing relations, it helped publicize the recuperation of the Lípez *ayllus* and the legal claim to territory its leaders were making. It may also have been a shrewd political move by the *cacique principal* (leader) to avoid any personal blame for the disaster. The quotation used above was gathered by a member of the NGO providing legal assistance for the land claim, but the herder's words encapsulate Lipeño worries. The failure to perform a *ch'alla* (ritual offering) to the high mountains, was responsible for the suffering wrought by the snowfall. The herder's words suggest an entanglement of people, land, animals and weather, and the need for Lipeños to acknowledge, maintain and renew their relation with the land. Renewing the ascent of Bonete, a mountain that retains snow on its uppermost slopes for most of the year, not only underlines human connections with the mountain but also their place in the weather world.

7 Conclusions

In July 2018, as I wrote the initial draft of this chapter, a Lipeño friend posted images on Facebook of a further snowfall in the province. The images were followed by comments of concern about the wellbeing of domestic animals. It would be hasty to conclude that extreme weather events are becoming more common – as scientists remind us, the recent occurrences could well be a temporary fluctuation in a predominantly stable pattern. However, global temperature rises will affect the water temperature in the Pacific Ocean, which drives the ENSO, to which extreme weather events in South America are linked. The present concern with weather and climate change makes Ingold's turn to the weather-world in anthropology an apt way of thinking through events like the snowfall of 2002. Likewise, Anderson's spatial focus on domestication – on the convivial places desired both by humans and animals – forces us to think about what happens when these places change. Together, in a changed economic conjuncture as well as a time of climatic flux, they compel us to ponder the responses, resilience and vulnerabilities of camelid herders and their herds.

Lipeño herders and their animals have together found and forged convivial spaces in the land and atmosphere of their marginal province. The normal predictability of the weather has enabled them to adopt non-labour-intensive techniques of herd management, which has in turn enables humans to adapt to changed economic circumstances. However, the absence of family members means that at times of crisis, fewer people are available to care for animals. In addition, herders experience extreme weather eventsas increasing in frequency. Domestication is a fragile arrangement, and the snowfall of 2002 shows us how the weather not only unites humans, animals, land and water, but can also separate them, rendering the normally convivial places of domestication hostile and dangerous. While some worry that migration is threatening rural lifestyles in the Andes, Lipeño people have adapted their herding to cope with new economic pressures. However, it remains to be seen whether their herding lifestyle can simultaneously withstand the additional pressures from climate change and inclement weather.

References

Allen, C. J. (1988). *The hold life has: Coca and cultural identity in an Andean community.* Smithsonian.

Anderson, D.G. (2014). Cultures of reciprocity and cultures of control in the circumpolar north. *Journal of Northern Studies*, 8(2), 11–27.

Anderson, D.G., Loovers, J.P., Schroer, S.A., & R.P. Wishart (2017). Architectures of domestication: On emplacing human-animal relations in the north. *Journal of the Royal Anthropological Institute*, 23(2), 398–416. DOI: 10.1111/1467-9655.12613.

Bastien, J. (1985). Qollahuaya-Andean body concepts: A topographical-hydraulic model of physiology. *American Anthropologist*, 87(3), 595–611.

Betanzos, J. de (1987) [1557]. *Suma y narración de los Incas*. Ediciones Atlas.

Bold, R. (2019). Contamination, climate change, and cosmopolitical resonance in Kaata, Bolivia. In R. Bold (ed.), *Indigenous perceptions of the end of the world.* Palgrave Macmillan.

Bolton, M. (2000). *Between the ayllu and the nation-state: Intertextuality and ambiguities of identity in San Pablo de Lípez*. PhD thesis, University of St. Andrews. https://research-repository.st-andrews.ac.uk/handle/10023/2005.

Bolton, M. (2002). Doing *waki* in San Pablo de Lípez: Reciprocity between the living and the dead. *Anthropos*, 97, 379–96.

Bolton, M. (2006a). The Devil and María Picha Picha at San Antonio del Nuevo Mundo: A *pachakuti* in the colonial past of a mining area. In P. Dransart (Ed.), *Kay pacha: Cultivating earth and water in the Andes* (pp. 47–54). Archaeopress, BAR International Series 1478.

Bolton, M. (2006b). Genetic defects or generative prototypes? Competing models for livestock improvement in southern Bolivia. *Journal of the Royal Anthropological Institute,* 12(3), 531-549.

Bolton, M. (2010). Healthy meat and natural animals: Nature, culture and decolonising knowledge in Bolivia. In M. Bolton & C. Degnen (Eds.). *Animals and science: From colonial encounters to the biotech industry* (pp. 169–195). Cambridge Scholars Publishing.

Bugallo, L., Dransart, P., & Pazzarelli, F. (Eds.), (2022). *Animales humanos, humanos animales: Relaciones y transformaciones en mundos indígenas sudamericanos*. Editorial Antropofagia.

de la Cadena, M. (2014). *Runa*: Human but not only. *Hau: Journal of Ethnographic Theory*, 4(2), 253–259.

de la Cadena, M. (2015). *Earth beings: Ecologies of practice across Andean worlds*. Duke University Press.

Capoche, L. (1959) [1585]. *Relación general del asiento y Villa Imperial de Potosí y de las cosas más importantes a su gobierno, dirigida al Excmo. Sr. Don Hernando de Torres y Portugal, conde del Villar y virrey del Perú*. Biblioteca de Autores Españoles, vol. 122. Ediciones Atlas.

Cassidy, R. & Mullin, M. (Eds.), (2007). *Where the wild things are now: Domestication reconsidered*. Berg.

Convery, I., Baily, C., Mort, M. & Baxter, J. (2005). Death in the wrong place? Emotional geographies of the UK 2001 foot and mouth disease epidemic. *Journal of Rural Studies*, 21, 99–109.

Cooperación Alemana al Desarrollo. (2003). *Proyecto emergencia nevada 2002: Informe final de la ayuda alemana a los afectados por la nevada de julio de 2002 en las regiones de los Lípez y Sajama*. PGR, SERNAP, GTZ.

Despret, V. (2004). The body we care for: Figures of anthropo-zoo-genesis. *Body and Society*, 10 (2–3), 111–134.

Dransart, P. (2002). *Earth, water, fleece and fabric: An ethnography and archaeology of Andean camelid herding*. Routledge.

Dransart, P. (2010). Animals and their possessions. In M. Bolton & C. Degnen (Eds.), *Animals and science: From colonial encounters to the biotech industry* (pp. 84–104). Cambridge Scholars Publications.

FAO Bolivia. (2009). *Identificación de indicadores y estados de alerta para el pronónstico y respuesta antes nevada, granizada, helada y riadas en Potosí*. Retrieved July 18 2018, from: https://issuu.com/ucerbolivia/docs/rrd_identificaci__n_y_estado_de_ale.

Gil Montero, R. (2014). *Ciudades efímeras. El ciclo minero de la plata en Lípez (Bolivia), siglos XVI–XIX*. Plural/IFEA.

Harraway, D. (2008). *When species meet*. University of Minnesota Press.

Harris, O. (2000). The power of signs: Gender, culture and the wild. In *To make the earth bear fruit: Ethnographic essays on fertility. work and gender in highland Bolivia* (pp. 180–200). Institute of Latin American Studies.

Harris, O. (n.d.). *The history of ayllus in the present conjuncture: Source of insight or waste of time*. Unpublished manuscript.

Ingold, T. (1994). From trust to domination: An alternative history of human-animal relations. In A. Manning & J. Serpel (Eds.), *Animals in Human Society: Changing Perspectives* (pp. 1–22). Routledge.

Ingold, T. (2000). *The perception of the environment: Essays in livelihood, dwelling and skill*. Routledge.

Ingold, T. (2007). Earth, sky, wind and weather. *Journal of the Royal Anthropological Institute*, 13 (Special Issue). S19–S38.

Ingold, T. (2010). Footprints through the weather-world: walking, breathing, knowing. *Journal of the Royal Anthropological Institute*, 16 (Special Issue), S121–S139.

Ingold, T. (2011). *Being alive: Essays on movement, knowledge and description*. Routledge.

Latour, B. (1993). *We have never been modern*. Harvester.

Lestel, D., Brunois, F. & Gaunet, F. (2006). Etho-ethnology and ethno-ethology. *Social Science Information*, 45, 155–177.

Lozano Machuca, J. (1992) [1581]. Carta del factor de Potosí Juan Lozano Machuca (al virrey el Perú Don Martín Enriquez) en que cuenta de cosas de aquella villa y de las minas de los Lípes (Año 1581). Transcribed by J.M. Cassassas C. *Estudios Atacameños*, 10, 30–34.

Lowe, P. (2004). *Ayllus originarios de sud oeste Potosí*. ISALP.

Martínez, G. (1983). Los dioses de los cerros en los Andes. *Journal de la Société des Américanistes*, 69, 85–115.

Martínez, G. (1989) *Espacio y pensamiento I: Andes meridionales*. HISBOL.

Nielsen, A. (1997–98). Tráfico de caravanas en el sur de Bolivia: Observaciones etnográficas e implicancias arqueológicas, *Relaciones de la Sociedad Argentina de Antropología* XXII–XXIII, 139–178.

Nielsen, A. (2009). Pastoralism and the non-pastoral world in the late pre-Colombian southern Andes (1000–1535). *Nomadic Peoples*, 13(2), 17–35.

Ødegaard, C.V. (2011). Sources of danger and prosperity in the Peruvian Andes: Mobility in a powerful landscape. *Journal of the Royal Anthropological Institute*, 17(2), 339–355.

Pazzarelli, F. (2022). Animales Paralelos: dualismos y perspectivas en los Andes meridionales. In L. Bugallo, P. Dransart and F. Pazzarelli (Eds.), *Animales humanos, humanos animales: Relaciones y transformaciones en mundos sudamericanos* (pp. 299–325). Editorial Antropofagia.

Platt, T. (1987). Calendarios tributarios e intervención mercantil: Racionalidades estacionales entre los indios de Lípez (Bolivia) en el Siglo XIX. In B. Larson, O. Harris & E. Tandeter (Eds.), *La participación indigena en los mercados surandinos: Estrategias y reproducción social, siglos XVI–XX* (pp. 471–557). CERES.

Platt, T. (2002). El feto agresivo. Parto, formación de la persona y mito-historia en los Andes. *Anuario de estudios americanos*, 58(2), 633–78.

Salomon, F. (1991). Introductory essay: The Huarochiri manuscript. In F. Salomon & G. Urioste (Eds.), *The Huarochirí manuscript: A testament of ancient and colonial Andean religion* (pp. 1–42). University of Texas Press.

Sillar, B. (2009). The social agency of things? Animism and materiality in the Andes. *Cambridge Archaeological Journal*, 19(3), 367–377.

Stépanoff, C. (2017). Comment on Anderson et al., Architectures of domestication. *Journal of the Royal Anthropological Institute*, 23(2), 417–418.

CHAPTER 7

'They Work for Me, I Work for Them': Investigatory Attunements and Partnerships between Dogs and Gwich'in in Northern Canada

Jan Peter Laurens Loovers and Robert P. Wishart

Abstract

In this chapter we explore the notion of investigatory attunement between Gwich'in dogmushers Abraham Stewart Junior and Joseph Kaye. Following their teachings and lives with dogs, we are careful to consider dogs as 'production of work' and instead advocate the notion of partnership to underscore the entanglements between dogs, Gwich'in, fish and other beings like the crow. In similar vein we are hesitant to consider these partnerships as social construction or as conventional understandings of domestication which entail forms of control or domination without any sense of mutual benefit. What the two Gwich'in dog mushers illustrate what living with dogs entails in the North is a more complex form of relationship that includes other beings such as fish within this life-world.

Keywords

Gwich'in – dogs – fish – partnership – domestication – work – attunement – Arctic

1 Introduction

> 'There was this old timer, a really good hunter and trapper, all summer he would catch fish for his dogs but he also gave some to crow.[1] He would always do that, cut the fish up into small pieces so the crow could eat it easily. Those crows they would help him out too. After freeze up he would travel with dogs and he still fed some of that fish to crow, meat

1 'Crow' is the Gwich'in English gloss for raven (*Corvus corax*) in Fort McPherson. *Deetr'in* in Teetl'it Gwich'in language.

 | DOI:10.1163/9789004679450_008

> too sometimes. Those crows would fly ahead of his dogs and watch out, if they saw caribou or maybe moose they would circle over them and he would go right there and he would be happy, the dogs would be happy, crows too I guess'. A few days after recounting this version of the story the elder expanded upon it, again recounting how the old timer would 'work hard' and feed both the dogs and crow and how the dogs would help him out, they were like his partners. He ended by recounting how the crows participated in this relationship further by pointing out that they too would help him. At one time, 'the man was travelling by boat and the kicker [outboard motor] started to just quit on him, he took the boat to shore and started to work on the kicker but gee he couldn't figure out the problem. Those crows they sat up in that tree watching him and finally called out: spleerk pluug, spleerk pluug. They were right, he cleaned that spark plug up and took off out of there'.

The above vignette from Wishart's fieldnotes developed out of making inquiries about the importance of fishing for the provisioning of dog teams in northern Canada. In this story Thomas Koe illustrates how a fixation on hunting, or fishing, or trapping as separate entities is difficult to reconcile with an understanding of 'working' that positions these as parts of a whole. In a similar vein, the repositioning of animals like crows into these extended life worlds by our Gwich'in teachers reminds us that while we may be fixated on dogs, fish, or caribou there are multiple other persons of concern. In this paper we want to do justice to what Gwich'in say regarding all the work that they and the animals with whom they form relationships do in a world that is called 'Gwich'in country' – a place continually worked and reshaped by the interactions of these various relations. This work entails developing partnerships and considerable interplay between humans and animals and animals amongst themselves. Despite recent rich theoretical accounts of human-animal relationships in post-humanist literature and the application of *perspectivism* and STS/ANT, these approaches do not give complete justice to Gwich'in life ways. Instead, we focus on a periphery of domestication where people are not *enacting with* animals (as STS literature advocates) but *working with* them. These working partnerships involve investigatory attunements (c.f. Shapiro 1990). Dogs become our illustrative point, a domesticate in a world of supposed wilderness. However, to set the scene, we must first explore how certain recent anthropological theoretical fixations have become almost gospel and then set the political and legal landscape in order to move into a world of valued partnerships.

2 Perspectivism

Viveiros de Castro's notion of 'perspectivism' (1998, 2012) or 'perspectival multinaturalism' (2004) has been widely used in anthropology to understand the complexities of animals, spirits and Indigenous Peoples (e.g. Brightman, Grotti & Ulturgasheva, 2012; Willerslev, 2007). His starting point is that within the Amazonian Indigenous world, animals were persons in mythical times and continue to share this capacity. Breaking with Darwinian evolutionary approaches that take animality as the starting point for human uniqueness, this world view, particularly as represented through shamanic philosophy, presents challenges to modernist thinking and the assumptions found within the nature-culture divide. Moving onwards, he elaborates on the notions of 'animism' and 'multinaturalism'. With 'multinaturalism', Viveiros de Castro reverses the relativist notion of multiple cultures and one Nature and implies that there are multiple natures but only one culture – a 'spiritual unity and a corporeal diversity' (1998, p. 470; 2004, p. 466). Taking inspiration from Descola, Viveiros de Castro redefines animism as 'an ontology which postulates the social character of relations between humans and non-humans' (1998, p. 473). The understanding of these relations is made possible through a particular point of view that then shapes or creates the subject (ibid, p. 476–7). As there are multiple points of view, there is a correlation between perspectivism and multinaturalism. These are not representations (e.g., an animal is like a human) but realities (e.g., an animal is a person from their own point of view) (see also Viveiros de Castro, 2012, p. 83). Picking up Viveiros de Castro's argument, Kohn argues that Ávila Runa, in the Ecuadorian Amazon, consider that 'all beings, and not just humans, engage with the world and with each other as selves – that is, as beings that have a point of view' (2013, p. 4). The subjective worldviews of sentient beings, however, are ontologically different but can be shared through inhabiting another world – or *umwelt* in his paraphrasing of Von Uexküll – and henceforth gaining another perspective (ibid, p. 7; for a dog *umwelt*, see also Horowitz and Hecht, 2014).

While the relational manner in which this argument tears down the boundaries between different selves is welcome, and Indigenous Peoples amongst others have been telling of their relations in ways that problematize Cartesian dualities, there is a problem of the generic to the specific. The problem emerges within anthropology in a curious and rather ironic way that is of no real fault or intention of the aforementioned authors. Turning the unique perspectives of Amazonian Indigenous Peoples into generic explanation of *all* Indigenous peoples, we fall into a similar trap as scientific naturalism which the authors tried to avoid. We question the tendency to consider

Viveiros de Castro's ontological take as the only and rightful way of understanding relations between Indigenous Peoples, animals, and spirits in the Americas and elsewhere. While not undermining or dismissing the importance of these ontological endeavours, we are mindful that they do not fit well with what our Gwich'in teachers and collaborators try to convey. The crows, in our starting vignette, were not perceived by the Gwich'in elder as a human-like person but rather as animals with whom he had established particular partnerships based on autonomy, intentions, care, and attunement. The crows, too, did not appear as dog-like persons to the dog but likewise as partners with whom they had formed a mutual contract (Wishart, 2003). Hence, we find the notion of *partnership* to be more fruitful in exploring human-animal relationships. As our vignette illustrates, dogs play a pivotal part in these relationships.

3 Partners at Work

DeMello describes dogs as 'man's first nonhuman partner' (DeMello 2012, p. 194). Zeuner (1963), likewise, argues that scavenging wolves or wild dogs came to live on the periphery of human settlements during pre-historical times. A co-operation between humans and these wild dogs in hunting other animals eventually led to the domestication of the animals. Zeuner goes on to say that the 'comradeship between man and wolf led to a driving of certain nomadic animals' (ibid, p. 61). While Zeuner provides a detailed overview of different dogs throughout history, the co-operation, or unequal symbiosis, is not further explored. Kohn provides some more insight and eloquently elaborates on the entanglements between dogs and humans. He states that through the 'mutual attempts to live together and make sense of each other, dogs and people increasingly come to partake in a shared constellation of attributes and dispositions – a sort of shared trans-species habitus' (2007, p. 7). This 'shared trans-species habitus' resonates with the work of Coppinger and Coppinger (2002) in behavioural studies and with M. Anderson's writing where Saami-herders consider their dogs as partners (1986, 2018). She goes to some length to elaborate on the ways of co-domestication and the particularities needed for dogs to be considered partners.

Northern Indigenous Peoples in Canada have made reference to partnership in several ways, though predominantly the term is related to human-human relations. Slobodin, working with Teetł'it Gwich'in in the 1940s, defines it as follows: "Partner' is a lifelong formal adoptive relationship between a pair of children or adolescents of the same sex. Partnership in this sense is far from universal among modern Peel River Kutchin' (1969, p. 87). Jarvenpa (1977) discusses the Denésoliné term *sits'eni* for partnership. Together with Brumbach (1997, 2006), Jarvenpa shows that *sits'eni* implicates a Denésoliné work team or unit on the symbolic level, but it can further 'imply friendship and reciprocal bonds that extend outside the domain of hunting and work' (2006, p. 42) where there are not apparent kin relations. Jarvenpa and Brumbach, however, do not extent this notion of *sits'eni* to Denésoliné-dog relations. Partnership from their perspective remains a human domain. Feit provides a different perspective. He speaks of 'mutual responsibilities and their abrogation or fulfilment' (2004, p. 99) in Néhinaw relations with animals and other humans, which does not necessarily imply egalitarian but rather respect and reciprocity. He elaborates that '[a]nimals are partners not only in the chase, but in the histories they produce' (ibid, p. 95). The idea of partnership between humans and dogs is also developed indirectly by Laugrand and Oosten. Resonating with some of our conversations with Gwich'in, they cite several Inuit who speak of the shared intimacy and entanglement – or symbiosis in their words – between them and their dogs. They go on to say that Inuit have several words to express such symbiosis; e.g., '*qimuksiit*' imples 'the collective action of dogs and humans when they are pulling the sled together' or '*angajuqqausiqpuq*' which refers to 'a person or dog trying to abuse its power or leadership position' (c.f. Therrien *in* Laugrand & Oosten, 2015, p. 154). The dogs are also considered to be 'protectors of the community' (ibid, p. 163–70) in which they can sacrifice themselves to take away illness or even death from a human. Analyzing socialisation between dogs and Sahtú, Savishinisky argues that 'the process of sensory and social extension is ultimately an act of mutual incorporation and learning' (1994, p. 216) in which dogs are 'social, sensory, and psychological, as well as an economic resource, and they are a self-producing resource at that' (ibid, p. 217). Savishinisky, like the symbiosis approach based on Zeuner, states that humans are as much domesticated by animals as vice versa.

All these partnerships involve work. Or, as Porcher defines the domestication of farm animals, living together entails a 'cooperative process of inserting animals into human society through work which involves, as Marx wrote, elements of exploitation and alienation, but also, and more particularly, the prospect of emancipation' (2014, p. 1). Coining the notion of 'interspecies solidarity' to investigate human-animal work relations, Coulter (2016) advocates an ethical and political framework in our dealings with animals as co-workers. Such solidarity breaks the boundaries between nature-culture, wild-domesticate, urban-rural (ibid, p. 12). Furthering ideas on animal work, Coulter (ibid, pp. 60–1) identifies three categories: subsistence work, voluntary work, and work mandated by humans. The first type is linked to the wild and hunting or trapping for subsistence. The work performed by animals might be debatable but human activities to get food is generally considered as work or labour. The second type concerns protective or caring work by animals for humans. Finally, the third type involves the group that includes industrial farming. The humans 'dominate' the work that animals perform. As we shall see below, working dogs in the Canadian North are difficult to place exclusively in any of these categories and rather are a blend. All these types, however, involve bodies that work. Following Wolkowitz, Coulter considers 'body work' as an analytical venue to investigate how different bodies mix. Or, as Coulter writes about equine work, '[t]he body work people perform on themselves makes it possible for them to work on/with horses' (2016, p. 31). Work, crudely said, always involves bodies as 'materials of production' (ibid, p. 31, 74). Elsewhere Loovers (2015) has written about 'dog craft', in which both dog-related artefacts and dogs are crafted in accordance with expanding (or declining) capitalist markets. Both 'body work' and 'dog craft' concern how animal bodies are made, trained, and maintained for particular tasks. Simultaneously, such crafting affects human bodies. This should not be mistaken as a rosy and harmonious attunement to the other body. Coulter is quick to note that body work, or interspecies work, involves resistance, manipulation, force, kindness, persuasion, cleanness and dirtiness, and care (2016, pp. 32–33, 71–95).

4 Gwich'in and Dogs

Much akin to Coulter's 'interspecies solidarity', a common phrase that we have heard from a number of Gwich'in people has been 'they work for us, we work for them'. With 'they' and 'them', the Gwich'in are referring to their former working dogs that were pivotal in the movement on the land and sustaining life. The idea of work might conjure an immediate association with capitalist modes of production but caution is warranted as Gwich'in reference to the word 'work' can imply other things. Indeed, while people reflect upon their days travelling with dogs and fishing for dogs they would often say that it was 'hard work', before arguing that at that time they did not consider it as work. In order to understand the phrase 'they work for us, we work for them', we argue that the aforesaid notion of partnership is useful. Here we note, as Coulter does (2016, p. 142), that partnerships do not exist in a vacuum from broader processes in life such as settler histories and colonialisation.

We would like to turn in the next sections to the working dogs of the Gwich'in. The dogs immediately demonstrate a stubborn refusal to fit neatly within any theory of domination and symbiosis (see also Anderson, Loovers, Schroer & Wishart, 2017). Part of the reason is the historical constitution of dogs and other animals in the Gwich'in world, and part of the problem rests with the idea of partnership rather than domination. Following conversations with Teetł'it Gwich'in we can make a generic distinction between different kinds of dogs: namely, (i) 'working' or 'freight dogs'; (ii) 'racing dogs', (iii) 'guard or camp dogs', and (iv) 'pet dogs'. Through historical research we might add also the 'pack/guard dog', the so-called Hare Indian Dog (Cummins, 2002), that existed prior to the arrival of Hudson's Bay Company working dogs (see also Loovers, 2015; McCormack, 2018). While the first three entanglements as well as the earlier dogs, could be understood as partners in various degrees, the 'pet dog' seems to be much like Haraway's companion (2003, 2008). An important difference, besides size and activities, is that 'pet dogs' have been living inside the homes of Gwich'in while the other dogs have been predominantly living outside.[2] At the same time, 'pet dogs' are often tied up on a short leash and even are taken for a walk on a leash. More recently, it has not been uncommon to see 'pet dogs' roaming around town. For the purpose of this article, however, our main focus is on working dogs with some references to 'race dogs' and 'pet dogs'. As illustration let us share some of the stories relayed by two Gwich'in men.

4.1 *Fishing with Babe, Caring for Dogs*

Abraham Stewart Junior ('Abe Junior') has been one of the very few, if not the only person, in Fort McPherson who has started up a dog team again with working dogs after a period of abandonment (see Figure 7.1; see also Loovers, 2015, 2018). During two lengthy interviews at his house in Fort McPherson, in early June 2013, Abe Junior spoke with Loovers about his life with dogs. The following articulation by Abe Junior provides a vivid example of a partnership and the making of an intimate relationship between Teetł'it Gwich'in and dogs:

> I had one leader, I called him Babe. … [W]hen I got him as a pup, you could not even touch him, he was so nervous. … You tie him up, and when you go there he just pull wind. I work with him. Every morning I go up

2 Dmitry Arzyutov has given a paper in St-Petersburg in which he brings together kinship theories with the way Nenet relate to dogs (personal communication). Unlike the Gwich'in, the Nenet have dogs in their tents and homes. We have a somewhat similar understanding of partnership, but these nuances of living with dogs differently are important to bear in mind and we prefer to take the route of partnership rather than kinship to discuss dog and human entanglements.

> to him, pet him. Pretty soon he was my best friend. I started [training/working with him], I want him for leader because I know he was really fast huh. You look for fast dog too, the one that will stay ahead. So, he was … a really fast learner. … [H]e just go any way you want him to go, and he stay ahead too. And we became close friends you know. One time we were down at Mouth of Peel way out on the Mackenzie and I set [fishnets under ice]. I had net in, two nets, and I was alone and had hard time to look at that [net]. I look at it and now I have to put it back in, I was having hard time. I took that dog his harness, tied it on to his harness and call him 'Babe', he listen to me, 'Okay Babe, go', he walk and I set net, tied [sic ??], I say 'Hoo' and he stop, he wait, and then I say 'Okay', and he go again. He set two nets for me. That was the kind of the dog he was (smiles) … Anyways he turned out to be real good dog, and he grew old. (Interview Abraham Stewart Junior, 6 June 2013).

This narrative of shared work with setting net is quite remarkable, and Abe Junior's comment about close friends is not often expressed so clearly, but it does not stand on its own. Apparent is that Babe understands Abe Junior's language as well as the shared task that is at hand: namely, setting the net under ice. The setting of nets under ice during the winter is a precise activity in which the proper rhythm, speed, and amount of pulling or not pulling are needed to have the fishnet set properly without being tangled up. It might not be too far-stretching to imagine Babe understanding that the fish caught in the net would be predominantly for him and his canine partners.

Abe Junior also speaks about caring for his dog: petting him in the morning. The next morning Loovers visited him again at his house to continue the interview. Recounting the story of Babe, he added that there was one man watching him setting net with Babe who had to laugh at seeing the dog pulling the net. Abe further elaborated on the aspects of caring and language:

> Some people they don't care [for their dogs], so … they don't have good dog team. It [the dogs] work but not very [much], you know for a couple of hours and then they are finished for the day. So, yeah them old people in the old days, they always had good dogs. That is because they talk to them, they talk to them like a human you know. I always tell my boys that, 'you know, when you go out to them dogs just go talk to them, just tell them anything'. I do that all the time. They gonna understand you at the end, I guess they start understanding you and when you want them to work for you, they gonna work for you. They wouldn't, they wouldn't, they are just willing to do what you tell them, because you [are] kind to them

> and they be kind to you. That is a dog. That is what I learnt from Old People like Simon Snowshoe and them. James Itsi too, he is a real good dog musher. Kind to dogs, he always had good dogs … all the time. (Interview Abraham Stewart Junior, 7 June 2013).

A few weeks prior to interviewing Abe Junior, Loovers had visited another Teetł'it Gwich'in man at his house. Joseph Kaye, who is a respected trapper in the community of Fort McPherson, provides a further nuanced perspective on dogs. The denominator of 'good dog' is fluid as he highlights in his account of trapping with dogs. After asking him about trapping with dogs, and stories that dogs know the locations of traps, Joseph answered in a somewhat surprising and illuminating fashion.

> Yes, oh gee, particularly when you going to go [trapping]… you know when you go to the lake and … set mink trap, along the shore. You'd be going along there and you'd just slow down and it'd make you sick too [that the dogs are resisting to work hard] … If your trap is always at the corner, they know it's there and it makes you mad. It makes you mad [that they are lazy and slow down] and you've got to scream at them till they go and when they look at trap they lie down and they know it too [that you are getting upset]. They know every trap and sometime the trap is not tighter and they slow down and it make them lazy.

FIGURE 7.1 Abraham Stewart Junior with his dog Oscar
PHOTO: JAN PETER LOOVERS, 2015

> [Peter Loovers] Why do they do that, I wonder?
>
> I don't know. They just do that ... That's why I say when you take dogs in the bush and when you go back to the same spot two or three years later they know the same spot, you know. And then you know where you'd set your mink trap, where they'd go slow. Especially when you are setting muskrat traps, then [the dogs] go lazy you know ... Always. Lots of times I used whip and hit the sleigh like that, but it's very tiresome. Tiresome. They get lazy and they know. Even good dogs, they got really good and then as soon as you start trapping then they get lazy. They get lazy too. But if you are on a long run they are okay. But as soon as they know muskrat trap, they know. And same as mink trap set, they know. If you go to Stoney [Creek] and to the mountain and to the cabin, no problem. They will just go [be travelling] good. But as soon as you start travelling they know. They know when to get lazy. Yup. They know. When you go to mountain [for hunting caribou] they just take off. As soon as you start setting muskrat trap, they get lazy and they know it [that they do not work as hard as they do when hunting or travelling up to the mountains].

Joseph Kaye's account resonates with Coulter's and Porcher's comments on animal work, body work, and resistance. Dogs do not always feel like working or doing particular tasks. Checking muskrat traps entails going from one trap to another often in close vicinity. The dogs, thus, only move for a little bit before stopping and within minutes moving again for a little bit.

4.2 *Dogs, Fish and Change*

Both Abe Junior and Joseph Kaye spoke of their lives with working dogs before the completion of the Dempster Highway in early 1980s. Once the Dempster Highway was constructed, the working dogs were 'given up' as transport and new kinds of dog breeds came into the community for different purposes: namely, racing dogs and pet dogs (Loovers 2015, 2018). Joseph Kaye was explicit in the difference of movement between racing dogs and working dogs:

> You see them [around town]. That's race dogs and not very [many] of them now. But it's different type of dog too. They gallop when they walk some. You know it's really ... funny for me. Because in the old days they trot but this one here is like galloping. But they are race dogs but I mean look around town and hardly anyone got dog teams. But that skidoo just destroyed everything. That spoiled everything. Everybody just lost interest in dogs. (Interview Joseph Kaye, 11 May 2013).

But it is not only the movement of the dogs that changed, the feeding and thus activities on the land have also changed. Once again, Joseph Kaye:

> [E]verything changed … because nobody got huskies now. Hardly any around … Everybody gave up on their dog teams. … If you have a dog team I guess you mostly feed them the same way. You give them white fish and fish. (Interview Joseph Kaye, 11 May 2013).

The feeding of fish resonates with Abe Junior's articulation of the amount of work on the land that involved feeding dogs rather than the amount of work that would involve feeding pet dogs:

> No, no you just never work hard enough. You got to fish, you got to fish and work hard [to feed dogs]. If you don't, you don't get what you need. … You can figure out how much you gonna need for the winter. And if you don't really work hard for it, then you have nothing. That is the law of our life that time. You had to work hard in order to succeed. It is still like that … them days. I hardly remember spending a dollar. You had to get your stuff out off the land. Just like today, we got all kinds of stores and you can buy anything you want. … So that is about my fishing there you know, *I fished, I fish, fish, fish.* (Interview Abraham Stewart Junior, 7 June 2013, our emphasis).

This seems to be in stark contrast with the feeding of dogs that currently takes place. Again, let us hear what Joseph Kaye has to say:

> Yeah, what I said, like everybody gave up their dogs and now you get them pet dogs and … they mostly like to eat store stuff. Weird, you know. A few times I have little dog around and I have a hard time to feed it. I tried to feed him fish but he just smell it and won't eat it. But as soon as he got crumbles he would eat it. (Interview Joseph Kaye, 11 May 2013).

Abe Junior found the same difference between feeding working dogs and 'pet dogs'. Asking him what changes he has seen in feedings dogs, he comments after a short contemplation:

> Yeah, well nowadays you can just [feed the dog anything]. If you throw scraps out there any dog is gonna eat it. I don't know what it is going to do to them but nothing I guess. But it don't matter because we don't [do] nothing with them, they are not working. It is just like me; I don't hardly work around here. I got hardly nothing to eat, it don't bother me I am not hungry. Dogs is same thing, but if they gonna work they gonna be hungry.

> I know that. That is the difference now, you just could give a dog anything and it [is] no problem; he is staying fat because he is not working. He is lying around all day long and some of them don't even walk around I guess. Yes. And that, that is a difference and you can go to store and buy those dog food, commercial dog food. I buy them once in a while, I give it to my dogs huh. But I don't like it because … that dog feed, that commercial [kind], it makes them shit all over. It is horrible, so I quit giving it to them. And now it is different, now I stop giving it to them. (Interview Abraham Stewart Junior, 6 June 2013).

Here Abe Junior connects food, work, well-being, people and dogs. Abe's personal experiences with feeding store-bought dog-food have not been positive and he describes the dog having diarrhoea. At the same time, he argues that dogs need good food (such as fish and caribou meat) from the land in order to be able to work. In a discussion about the history of Gwich'in country, an elder pointed out to Wishart that before the coming of mechanized transport 'Dogs [were] your power! And they [got] their power from fish!' Bringing fish into focus, the image of Gwich'in as hunter-gatherers living only in the moment – the cultural absolute zero – must give way to an image of multiple levels of cultivation. One elder put it like this to Wishart: 'You would not believe how much fish they bought from us. You know that HBC warehouse, biggest building in town back then. It was full to the rafters with bales of dryfish in early winter, then by spring it was full of furs. Everybody needed that fish not just us'.[3] The technology and practice of fishing developed by Gwich'in became a 'focal thing' (Strong & Higgs, 2000, p. 20), an appropriate way of doing things, that was adopted by the fur traders to create a successful way of developing the industry. The history of dry fish demonstrates a level at which fishing became integral to the fur trade as a source of fuel. But this mechanistic perspective is really only part of the story and the Gwich'in continue to place fish as a vital part of what living a good life comprises, despite the lessened need for them as a supply of dog feed.

5 Working Dogs in the North

DeMello illustrates how working dogs are a 'product of a very working environment' where animals are 'often seen as products or machines or units of value'

3 This statement was made in relation to a conversation about how many fish needed to be stored for people to be able to run the dog teams which made the trade in furs possible. He later estimated that a dog team would use 5000 fish over a winter's trapping season.

(2012, p. 219–20). These sentiments echo Descartes (2017 [1649]) who argued that beasts and animals are 'devoid of reason' and whose movements and passions are mechanical without intention or thought. DeMello elaborates that animals are socially constructed where the pet is a friend 'with history and interests and desires', while the working animal is produced for its economic value and is given away or killed after rendered useless. Menache (2000, p. 45) illustrates that Aristotle considered domesticated animals to be used for labour while wild animals were to be hunted. Similar to the biblical anthropocentric tendencies towards animals, Aristotle had also argued that animals were created to suit human purpose. Menache (ibid, p. 47–57) puts forth the argument that hunting dogs of medieval aristocratic families became mediators between the wild and the tame. She goes on to say that such partnerships between the nobility and their highly valued dogs did not eradicate the notion of human mastery over nature but it helped bring dogs into human society and to be reshaped as affective pets with particular laws and treatments. Treatises on communicating kindly to dogs were set in place by the nobility. The dogs of the nobility, thus, came to be beings with histories, desires, interests, and know-how. Arluke and Sanders consider the notions of 'wildness' and 'tameness' as socially constructed where the former implies 'distance and danger' and the latter the opposite (1996, p. 10). In similar vein, elsewhere the authors argue that animals have a physical being that is given a cultural identity by humans upon bringing them into the household, into civilisation (2009, p. XIII). They go on to say that the making of the dog as pet, or a family-member, begins with bringing the dog into the house. They call this the 'engrossment stage' (1996, p. 11; 2009, p. XIV), where pets are taught to obey the household rules while being made family-members. Such distinction is also argued by Savishinsky according to Menache (2000, p. 43) where the 'civilized' pets are anthropomorphized while those of the hunter are 'ephemeral members of their communities'.

To consider working dogs in 'Gwich'in country' as machines might be appealing. The dogs could be viewed as a replacement, or a prerequisite for 'mechanical vehicles' that can pull and haul loads of meat, fish, and fur across the land to different trading posts or communities. The dogs also could be considered like mechanical vehicles for transporting people. We are, however, careful about reducing dogs to such mechanized understandings. On the contrary, as the above examples have illustrated, dogs have intentions, intelligence, and sensitivities which are not executions of instinctive minds that carry forth mechanized bodily movements. Gwich'in do not make the dogs but rather, the partnership between Gwich'in and dogs illuminates the variable degrees of closeness and distance, knowing, and working together. Neither are these social constructs of dogs. Like Porcher (2017), we are also careful to consider partnership as an essentially harmonious relationship without moments of inequality. The breeding and feeding of dogs, for example, can entail some aspects of 'control', but we are reminded of D.G. Anderson's (2014) point of understanding needs and we would further this by considering these as *nuances in attunement*; what matters are the reciprocal or shared histories with dogs, Gwich'in, and other animals.

6 Conclusion

Recall the story of the crow with which we started this article, the Gwich'in hunter did not need to acquire the point of view of the crow to understand what it was saying in order to make sense of the provided clues neither did the hunter, crow, or caribou need to become an entanglement of 'humanness' to relate to each other. Rather, as Kohn (2013) points out, these are mutual attempts of living together. The crow did not perceive the problem with the outboard motor any differently from the hunter as it said, 'spleerk plug, spleerk plug'. Of course within the ontological turn, the idea of 'understanding' is being scrutinized. Building on radical alterity, what an 'other' thinks remains unknown to the anthropologist. What is left is that the 'other' provides concepts that can be used for anthropological purposes in order to 'to carry out an experiment with it' (Viveiros de Castro, 2013, p. 484). We are careful in taking forth such an 'ontological turn' per se. Considering the notion of partnership, thus, is not an experiment or an anthropological utilisation but rather a contemplation in relations between Gwich'in, dogs, crows, fish etc. Neither can it be said that the Gwich'in world is incommensurable with that of the West or of academia, indeed it would seem to be a problem of focus and vocabulary rather than the lack of shared common ground.

Shapiro offers an interesting alternative. While not undermining differentness between dogs and humans, he advocates a commonality through 'our mutually accessible bodily experiencing' (1990, p. 184). These shared experiences allows for a 'kinaesthetic empathy' (1990, p. 188). Whereas Shapiro takes a social constructive approach, like Menache and DeMello, and builds on von Uexküll's *umwelt*, like Kohn, we rather like to focus on how he exemplifies becoming and staying *attuned* to his dog Sabaka. Sketching the life history of Sabaka, as Abe Junior had done with Babe, Shapiro addresses how his life becomes (dis)entangled with that of his dog. He writes about his awareness of Sabaka's movements and anticipated his own bodily movements accordingly by living together. Or, as he writes, his own postures become investigatory as does Sabaka's (ibid, p. 190). This echoes Porcher (2017, p. 11), '[w]orking with animals is living with animals'. Partnerships, then, concern *investigatory attunements* where 'kinaesthetic empathy' allows working together in a more harmonious way while not dismissing moments of disconcert (Loovers, Losey

& Wishart, 2018). Abe Junior's account of setting nets with Babe is exemplary of a good partnership. Abe Junior and Joseph Kaye underscore how much work and sociality revolved around the procurement of fish to provide the fuel for various forms of Gwich'in partnerships, including the partnerships formed during the history of the fur trade (Loovers, 2015; Wishart, 2014). Or, as Abe Junior concluded, 'fish is always good for dogs, especially when they work huh. Because fish is always good for anything, anybody'.

Acknowledgements

We are grateful to Maggie Bolton and two anonymous reviewers who provided valuable guidance and feedback on an earlier version of this chapter. We are grateful for the ERC Arctic Domus grant (no. 295458). We acknowledge our Gwich'in teachers in Fort McPherson. In particular, Sharon Snowshoe, Arlyn Charlie, and Kristi Benson of the Gwich'in Tribal Council's Department of Culture and Heritage. This chapter was impossible without the generosity of Joseph Kaye and Abraham Stewart Junior to share with us their lives with dogs and fish. We dedicate this chapter to the late Thomas Koe and the late Abraham Stewart Junior.

References

Anderson, D.G. (2014). Cultures of reciprocity and cultures of control in the circumpolar north. *Journal of Northern Studies*, 9(2), 11–27.

Anderson, D.G., Loovers, J.P.L, Schroer, S.A. & Wishart, R.P. (2017). Architectures of domestication: On emplacing human-animal relations in the north. *Journal of the Royal Anthropological Institute*, 23(2), 398–416. DOI: 10.1111/1467-9655.12613.

Anderson, M. (1986). From predator to pet: Social relationships of the Saami reindeer-herding dog. *Central Issues in Anthropology*, 6(2), 3–11.

Anderson, M. (2018). Dogs in Saapmi. In R.J. Losey, R.P. Wishart & J.P.L. Loovers (Eds.), *Dogs in the North: Stories of cooperation and co-domestication* (pp. 267–277). Routledge.

Arluke, A. & Sanders, C. (1996). *Regarding animals*. Temple University Press.

Arluke, A. & Sanders, C. (2009). *Between the species: Readings in human-animal relations*. Pearson Education Inc.

Brightman, M., Grotti. V, & Ulturgasheva, O. (2012). *Animism in rainforest and tundra: Personhood, animals and things in contemporary Amazonia and Siberia*. Berghahn Books.

Brumbach, H.J. & Jarvenpa, R. (1997). Ethnoarchaeology of subsistence space and gender: A subarctic Dene case. *American Antiquity*, 62(3), 414–436.

Brumbach, H.J. & Jarvenpa, R. (2006). Chipewyan society and gender relations. In R. Jarvenpa & H.J. Brumbach (Eds.), *Circumpolar lives and livelihood: A comparitive ethnoarchaeology of gender and subsistence* (pp. 24–78). University of Nebraska Press.

Coppinger, R. & Coppinger, L. (2002). Dogs: A new understanding of canine origin, behaviour and evolution. University of Chicago Press.

Coulter, K. (2016). *Animals, work, and the promise of interspecies solidarity*. Palgrave Macmillan.

Cummins, B.D. (2002). *First Nations, first gogs: Canadian Aboriginal ethnocynology*. Detselig Enterprises Ltd.

DeMello, M. (2012). *Animals and society: An introduction to human-animal studies*. Columbia University Press.

Descartes, R. (2017 [1649]). *The passions of the soul*. Translated by Jonathan Bennett. Retrieved November 25 2018, from: https://www.earlymoderntexts.com/assets/pdfs/descartes1649part2.pdf.

Feit, H. (2004). James Bay Crees' life projects and politics: Histories of place, animal partners, and enduring relationships. In M. Blaser, H. Feit & G. McRae (Eds.), *In the way of development: Indigenous Peoples, life projects, and globalisation* (pp. 92–110). Zed Books.

Haraway, D. (2003). *The companion species manifesto: Dogs, people, and significant otherness*. Prickly Paradigm Press.

Haraway, D. (2008). *When speciesns meet*. University of Minnesota Press.

Horowitz, A., & Hecht, J. (2014). Looking at dogs: Moving from anthropocentrism to canid umwelt. In A. Horowitz (Ed.), *Domestic dog cognition and behaviour: The scientific study of Canis familiaris* (pp. 201–219). Springer-Verlag Publishing.

Jarvenpa, R. (1977). Subarctic Indian trappers and band society: The economics of male mobility. *Human Ecology*, 5(3), 223–259.

Kohn, E. (2007). How dogs dream: Amazonian natures and politics of transspecies engagements. *American Ethnologist*, 34(1), 3–24.

Kohn, E. (2013). *How forests think: Towards an anthropology beyond the human*. University of California Press.

Laugrand, F. & Oosten, J. (2015). The dog, partner of the hunter. In F. Laugrand & J. Oosten (Eds.), *Hunters, predators, and prey: Inuit perceptions of animals*, (pp. 151–178). New York and Berghahn Books.

Loovers, J.P.L. (2015). Dog craft: A history of Gwich'in and dogs in the Canadian north. *Hunter Gatherer Research*, 1(4), 387–419.

Loovers, J.P.L. (2018). 'Hard times are coming': Indeterminancy, prophecies, apocalypse, and dogs. In R.J. Losey, R.P. Wishart & J.P.L. Loovers (Eds.), *Dogs in the north: Stories of cooperation and co-domestication* (pp. 191–211). Routledge.

Loovers, J.P.L., R.J. Losey & R.P. Wishart (2018). Conclusion: Dogs in the north. In R.J. Losey, R.P. Wishart & J.P.L. Loovers (Eds.), *Dogs in the north: Stories of cooperation and co-domestication* (pp. 278–292). Routledge.

McCormack, P.A. (2018). An ethnohistory of dogs in the Mackenzie Basin (Western Subarctic). In R.J. Losey, R.P. Wishart & J.P.L. Loovers (Eds.), *Dogs in the north: Stories of cooperation and co-domestication* (pp. 105–151). Routledge.

Menache, S. (2000). Hunting and attachment to dogs in the pre-modern period. In A. Podberscek, E. Paul & J. Serpell (Eds.), *Companion animals and us: Exploring the relationships between people and pets* (pp. 42–60). Cambridge University Press.

Porcher, J. (2014). The work of animals: A challenge for social sciences. *Humanimalia,* 6(1). Retrieved November 25, 2018, from https://www.depauw.edu/humanimalia/issue%2011/porcher.html.

Porcher, J. (2017). *The ethics of animal labour: A collaborative utopia.* Palgrave MacMillan.

Savishinisky, J. (1994). *The trail of the hare: Environment and stress in a sub-Arctic community.* 2nd edition. Gordon and Breach.

Shapiro, K. (1990). Understanding dogs through kinesthetic empathy, social construction, and history. *Anthrozoös,* 3(3), 184–195.

Slobodin, R. (1969). *Leadership and participation in a Kutchin trapping party.* National Museum of Canada (Bulletin 228).

Strong, D. & E. Higgs. (2000). Borgmann's philosophy of technology. In E. Higgs, A. Light & D. Strong (Eds), *Technology and the good life?* (pp. 19–37). The University of Chicago Press.

Viveiros de Castro, E. (1998). Cosmological deixis and Amerindian perspectivism. *Journal of the Royal Anthropological Institute,* 4(3): 469–488.

Viveiros de Castro, E. (2004). Perspectival anthropology and the method of controlled equivocation. *Tipití: Journal of the Society for the Anthropology of Lowland South America,* 2(1): 3–22.

Viveiros de Castro, E. (2012). *Cosmological perspectivism in Amazonia and elsewhere.* HAU: Master Class Series, Vol 1.

Viveiros de Castro, E. (2013). The relative native. *HAU: Journal of Ethnographic Theory,* 3(3): 473–502.

Willerslev, R. (2007). *Soul hunters: Hunting, animism, and personhood among the Siberian Yukaghir.* University of California Press.

Wishart, R.P. (2003). *Living 'on the land': Teetl'it Gwich'in perspectives on continuities.* PhD thesis. University of Alberta.

Wishart, R.P. (2014). 'We ate lots of fish back then': The forgotten importance of fishing in Gwich'in country. *Polar Record,* 50(255), 343–353. DOI: 10.1017/S0032247413000715.

Zeuner, F.E. (1963). *A history of domesticated animals.* Hutchinson & Co LTD.

Afterword: Concepts that Travel

David G. Anderson

At the heart of this collection lies the inspiration that certain types of ecological relationships make themselves felt in vastly different parts of the world. At first glance, there can be nothing more different than the tropical green canopies of Amazonia, the towering peaks of the Andes, and the wide-open expanses of Arctic tundra. Yet these seemingly underdeveloped and sparsely populated regions enjoy a homology of intuition to which humans and non-humans relate. The implied irony is of course a sensory illusion. The Arctic, Andes and Amazonia are today, as at the time when ethnology was founded, at the forefront of global resource extraction. These incongruous chimeras are both isolated and crowded, pristine and planted. Within the cacophony of opposites we can find a unity of thought.

The red line that connects them are the 'sentient entanglements' symbolized by the meeting between the hunter and the hunted. The ethnographies that comprise this volume document variously charged partial connections between observant and empathetic beings (Strathern, 1991). The identity of these beings is poorly described by the words 'human' and 'animal' since in almost every case there is a shared circumstance that can be described as a partnership (Loovers & Wishart), a pilgrimage (Dransart), or a way of becoming (Vander Velden). The connections that bring these binary opposites together are often sustaining – such as the bodies offered to renew life. But they are sometimes confrontations, with deaths and killing providing a sombre backdrop that lends value to continued life.

These sentient entanglements come from a long line of travelling concepts, as Loovers & Bolton outline in their insightful introduction. The key term 'entanglement' finds its roots in the Scandic words for sea algae – and carries a connotation of an impossible intertwining of beings that can only be separated into specific opposing strands with great struggle. Sentience has its own history. Invigorated today with its connection to artificial intelligence, it has been a remarkably cosmopolitan concept over hundreds of years, governing intuitions of commensurability and connection among peoples and regions. The anthropologists Irving Hallowell and Tim Ingold are by far the most often quoted authorities in this volume on the commensurability of people and animals – of human and non-human persons. However, their texts sit upon readings of ethnography and philosophy that are quite old – and which point

 | DOI:10.1163/9789004679450_009

sometimes in diverging directions. Some of those older ideas are less entangled, more slippery, and often submerged.

Marylyn Strathern (2018), in a self-reflective chapter on her work on partible personhood, drew our attention to 'concepts that travel'. At the heart of her reflection was a somewhat sad puzzlement over what we might say was the 'viral' influence of her concept of the 'dividual' (Strathern, 1988) in comparison to that of the 'partible person'. The dividual travelled far and wide from India, to America, to Southeast Asia drawing attention to the fragmented yet connected nature of the self – while what she saw clearly as a related or commensurable concept was 'less tractable', and if you like 'entangled' in older dualisms which made it less mobile and somewhat parochial. Without taking sides on how best to define relational personhood, there is a similar commensurability contest at play with Hallowell's (1960, 2010) idea of the non-human person.

As Dransart notes in her contribution, Hallowell framed his Ojibwa [Anishinaabe] ethnography as conversation with consultants in an isolated hunting society with little exposure to Western modes of thought. The protagonists in his hard-to-find chapter can likely be cited by most anthropology undergraduates (the rock-that speaks [unlike those which are silent], thunder [with poor articulation], and the winds carrying messages). These are the paradoxical sentient entities which an urban reader would otherwise assume to be inanimate. Hallowell goes on to explain that the grammatical rules of the language itself assign agency to all objects in a manner that makes it impossible to think of an environment that is not always moving, watching, and reacting. Overwhelmed by these environmentalist examples it is easy to overlook the fact, that Hallowell himself notes, that these articulate other-than-human personages are not actually known as such to Anishinaabes – but instead as *ätísso'kanak*, 'our grandfathers' (1960, p. 27; 2010, p. 542) – as Dransart herself notes. Jeff Kochan (2015), in a masterful re-reading of the encounters with these senior kin, builds a theory of scientific neighbourliness around them. He notes that the famous speaking rock was brought to the attention of the Anishinaabe elder John Duck by a settler farmer who had dug it up by accident while planting potatoes. Out of respect – neighbourliness – he delivered it to his friend, the ritual specialist who, perhaps with a sly grin, spoke with the rock and verified that it had not been part of his ritual pavilion. Re-reading this passage with Kochan's insight, one now wonders if the passage of the thunderbird who misspoke ('Did you hear what was said?' 'No … I didn't catch it') was perhaps an attempt to pull the leg of the young ethnographer in a good-natured way. Was Hallowell taking these examples too seriously? The stories remind one of the raven crowing out to the puzzled hunters to clean the 'speerk plug' in Loovers & Wishart – a humorous reminder that animals are equally invested as people in properly functioning machines. Re-thinking these famous episodes is not to dismiss them, but instead an opportunity to reflect on the relationships and connections portrayed.

As the walk-in role of the potato farmer illustrates, in Hallowell's day these were not particularly isolated First Nations communities. His collaborators were likely strong in their traditions but also quite competent actors within, and co-authors of, the institutions of the settler state. Within circumpolar ethnography Algonquin interlocuters have been some of the most influential political translators glossing 'property' for a complex ritual relationship with animal spirit-masters (Feit, 1991) or 'gardens' or 'banks' for productive ecological systems (Feit, 2001). The idea of a non-human person, likely co-authored by Hallowell with his Anishinaabe colleagues, also travels well and over time makes increasingly powerful political arguments to an ecologically-minded public. But it is not unimportant to remind ourselves that the articulate personages of Anishinaabe also are more than that. They are also elder kinsmen who share their thoughts with their own descendants. When the grandfathers perform the role of an other-than-human person they are providing a partial connection to other worlds –a way of understanding and respecting our planetary neighbours. We likely cannot perform politics without neighbourly concepts, but we also need the measure of their limitations. The non-human person therefore is not so much an Ojibwa concept as a concept that points to peaceful relations between Canadian settler society and a knowledgeable Anishinaabe elders who regulate forest relationships with their grandfathers.

Equally as powerful, and well-travelled, has been the Anishinaabe concept of the *=doodem-* [stem for clan; extended family] which became refashioned into the world-renowned category of the totegain, as any anthropology undergraduate could likely recite, totemism refers to a human spiritual affinity or kinship to animal or plant with the memorable quality of being sacred or untouchable. The clearest cases, ironically, are not from Northern Canada but from the opposite side of the British Empire among Australian First Nations. The Aberdeen-based legal anthropologist John Ferguson McLellan (1869–70) drew together colonial-era travel reports from around the world – including European sources – to try to extract a common sentiment of affinity to the worship of animals and plants in all cultures. His Anishinaabe example, came second after George Grey's (2012, [1841]) description of the Northwestern Aborigine example of the *kobong* [friend or protector]. The Anishinaabe example was built around the very old account of the translator and trader John Long. Long's (1791, p. 86–7) account of *totamism*, which he served up as an example of 'the religious superstition of the Savages', is thoroughly dramatic. It describes the Shakespearean error that a hunter made in unknowingly shooting and killing a bear, which was his totem animal, after heeding a particularly prophetic dream. The hunter, it seems, was carrying a rifle, suggesting that already in the 18th century this society was also not particularly isolated. The hunter immediately fell down 'in a state of insensibility', and was later confronted by another bear who scratched his face and admonished him. The hunter was forced to explain himself to the bear, who was speaking on behalf of the animal masters. The emphasis in Long's account is on the hunter's over-reaction – and the account therefore has been dismissed for its ethnocentrism. McLellan himself selectively quotes from the original to build his own case for a cultural hierarchy of kinship forms, mother-right, and affinity with natural forms. His case for a universal human interest in natural actors irrespective of 'level' was an early attempt to describe a common human intuition which did not travel well. In the end, Emile Durkheim and Marcel Mauss (together with their students) re-packaged the totems within a complex universe of symbols and logical operators (Tarot, 1999). No one cites the slightly diverging path that John Long himself took. He defined *totamism* as 'this idea of destiny' (1791, p. 87).[1]

In my reading of Long's short ethnographic incident, destiny seems to me to be a quite different type of relationship than the totems that I was taught were 'good to think with' (Levi-Strauss, 1991). Long goes on to compare Chippewa *totams* to folk European beliefs that certain domestic animals often share the same fate as their owners, displaying illness or distress at the same time as their human twin. These kinsmen are entangled in a completely different sense, being part of a genealogical branch – they share a common fate. The idea of a prescribed destiny implies that hunters and bears, human persons and other-than-human persons, roosters and French peasants, travel the same path and share the same fortunes and misfortunes. They share a world, and do not enter it independently from stage right and then stage left to create a common scene. This idea of a fateful kinship embedded in a larger world reaches out to Bolton's evocation of the 'weather-world' into which all sentient entities are 'immersed'. Andean *ayllu* members, like those of a Saami *siida* (Sara, 2009) are understood to be the people, the domestic livestock and the sentient landscape all of which are united in a relationship of care and concern.

1 I am by no means competent in any First Nation language, but I understand that Long made a fateful innovation in playing with prefixes and suffices. *doodem* is a stem which cannot sit by itself without a personal prefix and possessive suffix: *odoodeman* 'him - his clan'. Long stripped the obligatory kinship affixes and instead inserted a political suffix – ism – which made the concept ready-to-travel (Department of American Indian Studies at the University of Minnesota 2002).

Long's *totamism* can certainly be said to have travelled well – in fact so much so that most people have forgotten from where it first departed. The idea of affinity with what many assume to be one's opposite (a non-human) is a very Christian metaphor which is well poised to moderate oppressive relationships on expanding frontiers – a role that John Long likely knew well. However, the original idea – the one that didn't travel – was instead that of a shared trajectory. This is neither an embodied perspectivism (Viveiros De Castro, 1998) nor a collapsing dualism (Ingold, 2000). Humans with their kinsmen animals, be they settlers or aboriginal inhabitants, likely all seek ways to live well to achieve their destinies. In a disrupted political ecology of resource extraction and trade, a *totamism* which reconciles may be a necessary concept. However, from the point of view of those who inhabit the same space, a shared life-project – or destiny – is likely also important, and not (as Long implies) a neurotic superstition. This idea of a neighbourly project – a project where a potato planter can share an interest in a rock with a Medicine Man – does not seem to have travelled as well as it should have. The papers in this volume I suggest contribute to this project.

One of the most ambitious expressions of this project can be found in Vander Velden's chapter on Brazilian *onças* – a predatory cat which exists in a greater variety of forms among indigenous Karitianas than with biologists. The study of alternate systematics is a classic field – and it comes to the heart of the question of what ideas can travel. The idea that a designated species may come in behaviourally dangerous and benign forms, however, is not an idea that makes a soft landing in most biology departments. In terms of this volume the *onças* are exceptional since they are not thought to serve as food, unlike fish, caribou, or camelids. They have more in common with the mechanic-crow, and the totem-bear, who stand outside of circuits of sustenance as hostile reviewers. *Onça*-ness can also be encountered in domestic pets who turn feral. They exist 'between the everyday and extraordinary'. Vander Velden`s interpretation is that the liminal, transgressive *onças* represent a 'becoming' – something on its way to a different often monstrous form. Like with wild and tame reindeer, they are distinguished not by their biology but by their behaviour. It is this aspect of journeying-as-being which I think connects this chapter to the general theme of destiny (although in this case the *onças* are not so much sharing a path, but rather blocking it).

Frédéric Laugrand enriches this systematic study with his cosmology of caribou hunting in the Canadian Eastern Arctic. His emphasis is on what he identifies as an Inuit aversion to reindeer herding approaches the idea of the undesirable and inedible from the opposite direction. His rich examples demonstrate that many Inuit find herded reindeer to be inedible because they are not allowed to choose their own places, and spend too much time in and around people, making their most intimate lives transparent. In this case, a little bit of extraordinary creates a nourishing species. He dismisses the superficial homology that many modernising authors have had, from Birket-Smith (1929) to Stefansson (1922), that the localized control over *Rangifer* would improve Inuit society. I think all of us would also be sceptical of the modernising project – but I can't resist remarking that reindeer pastoralism across Eurasia is also ambiguous both practically and cosmologically. Many Northern hunters had a narrowly defined reindeer pastoralism forced on them by adverse circumstance – be it the elimination of free-ranging wild *Rangifer* in Scandinavia to the collectivisation projects of the Soviets. Developing Takakura's (2010) model of concentric circular relationships, most Siberian peoples who choose to live with reindeer tend to develop a range of 'relationships of familiarity' with their animals. The most docile would not differ from the way that Simon Tookoome speaks of 'adoption'. These reindeer, often raised from birth by hand, stick around the camp and sometimes enter the tents. Some are thought to share a fate or destiny with their owners – and can be used in healing rituals (Anderson 2011). Other less habituated *Rangifer* nervously skirt the edge of the camp or may have to be brought back to camp constantly. They are often indistinguishable from wild reindeer, and often they are descended from wild stock. They are kept as a 'reindeer close to hand' – animals which can be, and must be, hunted as needed without having to travel far. Cosmologically, similar rituals of respect are performed to the land spirits to influence, perhaps indirectly, the reproduction of the 'outer' herd. These various dimensions of familiarity bear a closer resemblance to the ideas of kinship and destiny than to the stereotypes of domination. The lesson of these two chapters is that there are ranges of partial relationships – some sustaining, some threatening, but never a dualism that needs – re-entangling.

Carrying on with the study of partial connections, I find it useful to read Sautchuk's chapter together with that of Bolton. Although their sites are wildly different, I am captivated by their attention to the built environment and the tools, architectures and equipment used to help humans and non-humans to build relationships. Sautchuk broadly and ambitiously investigates 'forms of life and death' through comparison of the hunting of two very different types of fish. One fish – the *pirarucu* – must be individually harpooned in a difficult contest of intention and deception much like the hunted caribou or perhaps better a sea mammal. The yellow catfish *gurijuba* by contrast is gathered 'like Gold' and its prevalence reflects on the luck and destiny of the fisherman. Linking the different methods are techniques – a discussion well developed in the French anthropological literature and not so very often referenced in

English language studies of human-animal relations. Here techniques are bundles of *approaches* which go beyond a simple competence with a lasso or harpoon. They point to how relations of familiarity and destiny congeal into an encounter mediated by equipment that a hunter-fisher carries. The *gurijuba*, in particular, are said to take on characteristics similar to those of the fishermen in the sense that they amass around a properly maintained and operated boat, and around a skipper with good fortune. One is reminded of Barbara Bodenhorn's (1990) famous example that a successful whale hunt is not determined by the man who holds the harpoon but by the woman who sews his clothing. Here there is a sense of entanglement which helps us to define sentience. Bolton, with her ambitious application of Ingold's (2010) weather-world to Andean llama pastoralism, creates a wider context within which these techniques of encounter should be understood. Documenting cases of extreme weather events, which unravel the bonds between herders and llama, Bolton argues that people, animals, techniques and kit are all 'immersed' in the weather. As with Hallowell's classic examples, the Andean weather reacts to misfortunes and events within human/animal society. Citing some of my own co-authored work, she also shows how the instruments of domestication encourage mutual autonomy and communication but also define a setting where cooperation can occur. The weather-world, she suggests, should replace the affordances of the landscape – and I would agree heartily. However I would add, building on some of the examples we cited in Anderson, Loovers, Schroer, & Wishart (2017), that those home-like spaces of collaboration which smell of smoke and salt, and which are patrolled for predators, also often establish a microclimate that is independent of the larger shifting forces buffeting the weather out-there. Would that be a weather-encampment? I don't know much about llamas, but the use of smudge fires and shade-shelters to counteract the most annoying aspects of the heat (for unlike llamas, reindeer suffer most from heat rather than snow) the herders are able to bring a little bit of alpine down to the bottom of valleys – extending the weather out of its normal location. In so doing they encourage the animals to circle back to the camp making this otherwise stationary point a destination on a journey.

Penny Dransart in her chapter appeals to the language of personifications to encourage us to view domestic animals – llamas, alpacas and sheep (and perhaps motorized transport) – as pilgrims and wayfarers. She masterfully reviews the literature on animism, drawing strongly from Robert Brightman (2007), to argue that humans and animals do not existentially confront each other as opposites since animals on the whole do not find the human arts and skills either powerful or attractive – and they themselves have chosen a different path. In her words the 'axiomatic splits in nature-culture … do not cleave cleanly as they are supposed to do'. The *wayñu* fertility ritual, which forms the heart of her ethnography, brings together herdsmen, alpacas and llamas in an evening of song, re-affirming the shared path that all these living entities share – perhaps not a concept-that-travels but a travelling joint-commitment (Hanks 2020). Here entanglements are sentient in the sense that animals and human make choices to live together and to thrive together. Reading this chapter, I have to admit that I am curious also about the motorized vehicles that peasants use to weave in between the mountain ridges (which look like intestines) in order to affect their pilgrimages. What would be the limit of personifications in this case. I have seen Evenkis ritually smudging vehicles with juniper in the same way that not long ago they would smudge transport reindeer before entering a new watershed, and Bolton adds that vehicles in the Andes receive libations and sometimes blessings from a Catholic priest. If ravens can become mechanics, one wonders if vehicles can become pilgrims.

Loovers & Wishart beautifully bring all of these points home with their account of working dogs in Gwich'in country, Northern Canada. They take seriously the words of hunters who work with dogs that they and the dogs form a partnership by working together. They stand this in ironic juxtaposition to the literature that 'takes seriously' that people and animals 'enact' a relationship. The metaphor of working together in my opinion both travels well and also encapsulates some of the hidden meanings behind two hundred years of ethnography which tries to capture these relationships. In order to work well together, there must be a measure of understanding and a co-commitment to a project (as well as moments of misunderstanding and anger). As Loovers & Wishart note, Eastern Inuit, who in Laugrand's account shun partnerships with caribou, share many of the same sentiments with their working dogs. However unlike in Laugrand's account (or perhaps even that of Sautchuk), the extraordinary animal – the Crow who hints and taunts – is also somehow engaged in the same joint project to recreate life. As the authors note: 'the Gwich'in hunter did not need to acquire the point of view of the crow to understand what it was saying'. The common lived environment gave a context where both Crow and hunter could understand the dilemma of re-starting an engine.

When re-reading these chapters, how can we now place the notion of 'sentient entanglements' which gives voice to this collection? Through a re-reading of some not-often-consulted ethnography I have tried to argue that sentience draws its meaning from certain ideas which have not travelled-well. While human-animal commensurabilities draw upon attempts to index openly visible, public, qualities that link animals to human, sentient entanglements are often whispered among kinsmen and understood within special contexts. They tend to reflect choices and evaluations – positive in the case of the caribou which chooses a hunter; negative in the case of the trusted house pet which transforms into an *onça*. Re-reading accounts of Anishinaabe grandfathers and totem-animals, I am struck now by the evocativeness of how ideas of destiny and kinship encourage us to think of these entanglements in a different way – a way that means more than just seeing the human in the animal individual (or collapsing nature into culture). The etymology of entanglement is striking – its link to the way Scandics related to sea algae. In my own brief struggles with sea algae in Northern Norway, I would not immediately see *tang* [seaweed] as evocative as *ting* [assembly] in social theory. Aside from evenings spent recovering kelp from the high-tide line to enrich a potato plot, exposed *tang* was a hazard. I can still hear the voice exhorting me not to stand on the exposed sea algae and risk slipping into an unsettled sea. Sea algae causes problems, in my experience, and not in a good way. One can clearly imagine that kelp forests also create problems for skippers and fishermen – snagging fishing gear and wrapping themselves around propellors. The etymology clearly indicates struggle – with the dominant meaning of an entanglement being fist-fight or argument which is both unavoidable and impossible to separate. However, when not lying matted on the shore, sea algae tend to unfold into parallel and perhaps harmonic strands when flooded by the tide. In fact, one could not even describe them as any more tangled than the fir trees in a plantation. Among its stalks, cod smelts seek protection and find food. As ecologists argue in Northern Norway (Strånd 2019), restoring kelp after three decades of industrial fishing is the first step in encouraging many species to return to the fjords. As *onças* are to caribou, this is a type of entanglement that is life-affirming (if not equally inseparable). The literature on so-called human-animal relations has been invigorated for almost forty years by the project of patching-up binary oppositions. As each of the chapters in this volume affirm, perhaps now is the time to rejoice in the partial connections.

References

Anderson, D.G. (2011). Idioms of health and healing in central siberia, In David G. Anderson (Ed.), *The healing landscapes of central and southeastern siberia* (pp. 95–110). CCI.

Anderson, D.G., Loovers, J.P.L., Schroer, S.A., & Wishart, R.P. (2017). Architectures of domestication: On emplacing human-animal relations in the north. *Journal of the Royal Anthropological Institute,* 23 (2), 398–416.

Birket-Smith, K. (1929). *The caribou eskimos: Material and social life and their cultural position.* Gyldendalske boghandel.

Bodenhorn, B. (1990). I'm not the great hunter, my wife is: Inupiaq and anthropological models of gender. Études/Inuit/Studies, 14 (1–2), 55–74.

Brightman, R. (2007). Nature and culture in the bush: A nature/culture metalanguage and rock cree parallels, In F. Laugrand and J.G. Oosten (Eds.), *La nature des esprits dans les cosmologies autochtones* (pp. 31–44). Laval, Québec: Les Presses de l'Université Laval.

DepartmentofAmericanIndianStudiesattheUniversityofMinnesota.(2002).=doodem-, The Ojibwe People's Dictionary. https://ojibwe.lib.umn.edu/main-entry/doodem-nad.

Feit, H.A. (1991). The construction of algonquian hunting territories, In G. W. Stocking (Ed.), *Colonial situations: Essays in the contextualisation of ethnographic knowledge*, (pp. 109–134). University of Wisconsin Press.

Feit, H.A. (2001). Hunting, nature and metaphor: Political and discursive strategies in James Bay Cree resistance and autonomy. In J. A. Grim (Ed.), *Indigenous traditions and ecology* (pp. 411–452). Centre for the Study of World Religions, Harvard Divinity School and Harvard University Press.

Grey, G. (1841). *Journals of two expeditions of discovery in north-west and western Australia, during the years 1837, 38, and 39*. T and W Boone.

Grey, G. (2012). *Journals of two expeditions of discovery in north-west and western Australia, during the years 1837, 38, and 39*. Cambridge University Press.

Hallowell, A.I. (1960). Ojibwa ontology, behaviour, and world view, In S. Diamond (Ed.), *Culture in history: Essays in honour of paul radin* (pp. 17–49). New York: Columbia University Press.

Hallowell, A.I. (2010). Ojibwa ontology, behaviour and world view, In A.I. Hallowell, J.S.H. Brown, & S.E. Gray (Eds.), *Contributions to ojibwe studies: Essays, 1934–1972* (pp. 535–568). UNP – Nebraska Paperback.

Hanks, W. F. (2020). Joint commitment and common ground in a ritual event, In *Roots of human sociality* (pp. 299–328). Routledge.

Ingold, T. (2000). Totemism, animism and the depiction of animals, In *The perception of the environment: Essays in livelihood, dwelling and skill* (pp. 111–131). Routledge.

Ingoldí, T. (2010). Footprints through the weather-world: Walking, breathing, knowing. *Journal of the Royal Anthropological Institute* 16, S121–S139.

Kochan, J. (2015). Objective styles in northern field science. *Stud Hist Philos Sci*, 52, 1–12. DOI:10.1016/j.shpsa.2015.04.001.

Levi-Strauss, C. (1991). *Totemism*. Merlin Press.

Long, J. (1791). *Voyages and travels of an indian interpreter and trader, describing the manners and customs of the north american indians; with an account of the posts situated on the river saint laurence, lake ontario, &c. To which is added a vocabulary of the chippeway language. And a list of words in the iroquois, mohegan, shawanee, and esquimeaux tongues, and a table, shewing the analogy between the algonkin and chippeway languages.* Robson.

McLellan, J.F. (1869–70). The worship of animals and plants. *The Fortnightly Review* 4; 7: 6 [12] (1869), 407–27, 562–82; 7 [13] No. 37 (1870), 194–216.

Sara, M. N. (2009). Siida and traditional sami reindeer herding knowledge. *Northern Review* 30, 153–178.

Stefansson, V. (1922). *The northward course of empire ... With illustrations and a map*. G. G. Harrap & Co.

Strand, H.K. (2019). Porsangerfjorden 2.0—En mulighetsstudie. *Rapport fra havforskningen.*

Strathern, M. (1988). *The gender of the gift : Problems with women and problems with society in melanesia.* University of California Press.

Strathern, M. (1991). *Partial connections.* Rowman and Littlefield.

Strathern, M. (2018). Persons and partible persons, In M. Candea (Ed.), *Schools and styles of anthropological theory*, (pp. 236–46). Taylor & Francis.

Takakura, H. (2010). 'Arctic pastoralism in a subsistence continuum: A strategy for differentiating familiarity with animals', In F. Stammler & H. Takakura (Eds.), *Good to eat, good to live with: Nomads and animals in northern eurasia and africa* (pp. 21–42). Sendai, Centre for Northeast Asian Studies.

Tarot, C. (1999). *De Durkheim à Mauss. L'invention du symbolique : Sociologie et sciences des religions.* Découverte/M.A.U.S.S.

Viveiros De Castro, E. (1998). Cosmological deixis and amerindian perspectivism. *Journal of the Royal Anthropological Institute* 4 (3), 469–488.

Index

www.ingramcontent.com/pod-product-compliance
Lightning Source LLC
LaVergne TN
LVHW010626300926
849162LV00001B/84
* 9 7 8 9 0 0 4 6 7 9 4 4 3 *